From Genesis to Genetics

From Genesis to Genetics

The Case of Evolution and Creationism

John A. Moore

UNIVERSITY OF CALIFORNIA PRESS

Berkeley   Los Angeles   London

University of California Press
Berkeley and Los Angeles, California

University of California Press, Ltd.
London, England

First paperback printing 2003
© 2002 by The Regents of the University of California

Library of Congress Cataloging-in-Publication Data

Moore, John Alexander, 1915–
     From Genesis to genetics : the case of evolution and
creationism / John A. Moore.
          p. cm.
     Includes bibliographical references (p.).
     ISBN 0-520-24066-9 (pbk : alk. paper)
     1. Evolution (Biology)   2. Creationism.   I. Title.

  QH367 .M813 2002
  231.7'652—dc21

                              2001044419

Manufactured in the United States of America

11   10   09   08   07   06   05   04   03

10   9   8   7   6   5   4   3   2   1

The paper used in this publication meets the minimum
requirements of ANSI/NISO Z39.48-1992 (R 1997)
(*Permanence of Paper*). ∞

For Susan Wallace Boehmer,
whose warm personality, fine mind,
and firm editorial hand have made
a wonderful contribution to scholars
and scholarship

CONTENTS

FIGURES AND TABLES

FIGURES

TABLES

PREFACE

Of the three major conflicts between science and religion, two have already been settled. It is now generally accepted that the Earth is round, not flat, and that the Earth revolves around the sun, rather than the sun circling the Earth. The third major conflict concerns origins—the origins of the universe, the Earth, and all its living creatures. In Western culture until the middle of the nineteenth century the answer to the question of origins was divine creation as described in Genesis, the first book of the Judeo-Christian Bible. This explanation had satisfied most people in the West since the waning days of the Roman Empire.

But in 1859, with the publication of Charles Darwin's *On the Origin of Species,* an alternative explanation for the origin of life's diversity appeared, and this new view of life threw Western thought into a tailspin. Darwin proposed that the many different kinds of plants and animals we see around us have not been immutable since the beginning of time. Instead, they have changed dramatically as the environment has changed, dividing again and again into new species that fill new niches, until—over vast periods of time—a huge number of different species has come into being. Darwin called this process evolution.

The *Origin* immediately evoked an outcry in Darwin's own Great

Britain, as well as on the European continent and in the United States. Many people saw his radical theory as conflicting severely with the religious teachings of Genesis, as indeed it did—and still does. Divine creation is a supernatural concept that cannot be studied, proved, or disproved with the procedures of science. It is accepted on faith. In contrast, Darwin's explanation for the diversity of life relied solely on natural processes, and therefore it could be tested, modified, and improved upon.

As the nineteenth century progressed, more and more naturalists accepted evolution as an explanation for the great diversity of life because it cast so much light on a number of otherwise inexplicable puzzles in nature. Because naturalists study the life histories of plants and animals, it is not surprising that they were the first people to be persuaded by Mr. Darwin, a fellow naturalist. The various Christian denominations were much slower to appreciate his new theory, but by the second half of the twentieth century many of them also accepted evolution. And now, at the beginning of a new millennium, Pope John Paul II has decreed that evolution is consistent with the teachings of the Roman Catholic Church.

Still, a very large number of fundamentalist Christians continue to reject the scientific evidence for evolution and accept as fact the literal account of creation as described in Genesis. The tension between creationists and evolutionists has waxed and waned over the decades since Darwin first proposed his "dangerous idea," but it is once again severe. The arena of contention today is mainly the public schools. Scientists, supported by the nation's courts, demand that teachers present evolution—and only evolution—as the accepted scientific explanation for the origin of life's diversity. Creationists, on the other hand, demand that educators exclude evolution from the curriculum—but if educators must teach it, they should also be required to teach a competing theory, called creation science, as a "logical" alternative.

This bitter argument has disrupted science education in the nation's public schools. Many school boards, principals, and biology teachers have opted to ignore the topic of evolution altogether rather than face the anger of parents who argue that discussion of evolutionary concepts in the classroom is undermining their children's religious beliefs. Exasperated, some educators find themselves asking: "Does it really matter if children never learn about evolution?"

Yes, I believe that it matters a great deal, for much more than evolution is on trial. A very important principle is at stake—the principle that the science curriculum in the schools should reflect what scientists have accepted as the best that can be said about the natural world on the basis of the evidence available. The entire scientific enterprise rests on two important principles: (1) that scientists must minimize their personal preferences and beliefs about how the world works and, instead, base their conclusions on data from nature gathered through observation and experiment; and (2) that scientists must subject those observations, experiments, and conclusions to testing and confirmation by other scientists. This rigor is the very heart of the scientific effort. And yet our public schools seem to be in danger of jettisoning scientific procedures at just the moment when we need the tools of science most to help us solve a host of social and environmental problems. In fact, the data and methods of science must be used if humanity is to have a relatively benign future.

The material in this volume is for those who are concerned with these educational issues and with larger questions raised by the religion-versus-science debate, particularly the standoff between evolutionists and creationists. Thus, I write for parents who want what's best for their children, for science educators who regularly confront the seemingly intractable problem of what to teach and how to teach it, for scientists and religious leaders who face questions from concerned and disconcerted citizens, for civic leaders who recognize that a vigorous and effective scientific community is essential to our nation's health and to our leadership role in the larger world, and for all who desire harmony in a diverse society that daily seems to grow more contentious.

The chapters that follow lay out and evaluate the scientific evidence for and against creationism and evolution, as well as the consequences of adopting one or the other paradigm to explain the diversity of life. They also cover the history of this controversy and the patterns of thought that creationists and evolutionists evince. In the final chapter I explore some options for avoiding abrasive confrontations over this issue in the future, so that both science and religion can be free to do the good work for which each is uniquely qualified. I believe we can have peace—if peace is what we desire.

ACKNOWLEDGMENTS

In undertaking this major task, I have sought the advice of wise and informed individuals to help me say what is correct and what is useful. My wife, Dr. Betty C. Moore, edited the first draft of the manuscript. Those who have been most helpful with theology and religious history are Edwin S. Gaustad, Douglas M. Parrott, Raphael Zidovetzki (all of the University of California, Riverside), and the Reverend Cynthia P. Cain (late of the Universalist Unitarian Church of Riverside). Scientists, most of whom have long been concerned with the science and creationism problems, are Bruce Alberts (National Academy of Sciences), Andrew Ahlgren (American Association for the Advancement of Science), Francisco J. Ayala (University of California, Irvine), Charles Birch (Sydney University), Rodger Bybee (Biological Sciences Curriculum Study), Richard Cardullo (University of California, Riverside), Ursula Goodenough (Washington University), Arnold Grobman (BSCS and the University of Missouri, retired), Donald Kennedy (Stanford), Ernst Mayr (Harvard), Ingrith Deyrup Olsen (University of Washington), Clay Sassaman (University of California, Riverside), Eugenie Scott (National Center for Science Education), Claude Welsh (Macalester University, retired), and Rachael Wood (Delaware Department of Education). I also wish to thank two informed citizens,

Betsy Donley (Phoenix, Arizona) and Sally M. Gall (librettist, La Jolla, California) for their thoughtful comments. Help of a different but critically essential sort has come from Larry McGrath and his associates of the University of California, Riverside, Microcomputer Support Group.

It is customary not only to thank wonderful people like these but to say they are not responsible for any remaining errors. This is true, of course; but more important, there are many fewer errors and better arguments because of them. And their warm support and encouragement has made a very great difference in my attempts to deal with this very important, discouraging, and fascinating topic.

And then come the very special expressions of appreciation for those who are so indispensable in converting a manuscript into a printed book—the editors. Mine have been Susan Wallace Boehmer, Howard Boyer, Doris Kretschmer, Suzanne Knott, and Carolyn Bond. Their skill and encouragement have made all the difference.

ONE

When Worlds Collide

In July 1996 a nearly complete human skeleton was found by two college students on the bank of the Columbia River near the town of Kennewick, Washington. Human remains so encountered are always of concern—"Who was it?" and "Who did it?"—so the students called the police. In an effort to answer those two questions the police turned the bones over to the local coroner. Burial grounds of Native Americans are sometimes encountered in that part of the Northwest, and the Native American Graves Protection and Reparation Act, passed in 1990, required that any such remains be returned for burial to the tribe to which they belonged. The bones appeared to be of great age, and so it was assumed that the skeleton must be one of a Native American, since settlers of European origin reached the West Coast only a few centuries ago. But closer study suggested otherwise.

To solve the puzzle, the bones were examined by an anthropologist, James Chatters, a specialist in skeletal remains of human beings. Such professionals can determine sex, size, age, cause of death, and racial type with considerable accuracy. Examination showed the skeleton to be that of a 50-year-old male of medium build, his teeth well worn and a stone arrowhead imbedded in his hip bone. Radiometric methods determined that the man died about nine thousand years ago—

long before human beings of European origin first arrived in the New World, according to conventional historical accounts. Yet the skeleton had Caucasoid features. Chatters took the bones to another anthropologist for an opinion, without giving a hint of his analysis, and was told that the skeleton was of a Caucasian male. Even when Chatters revealed the age of the bones, the second anthropologist stuck to her original identification. A third anthropologist familiar with the skeletal features of modern tribes of Native Americans concluded that the skeleton could not be assigned to any one of them.

Finding the nearly nine-thousand-year-old skeleton of a Caucasoid male in any part of the New World is puzzling in the extreme. Traditionally, anthropologists have thought that the first human beings to inhabit the Western Hemisphere crossed from Siberia to Alaska about 15,000 years ago. Some now believe that the event occurred much earlier, but in any case these immigrants from Siberia were of the Mongolian racial type, as are Native Americans—not Caucasoids. Anthropologists know that a few nameless fishermen from Western Europe came to the eastern shores of the New World before Columbus's arrival in 1492, as did the Vikings, but Caucasians did not come in large numbers until early in the sixteenth century.

So who was the Caucasoid Kennewick Man who arrived thousands of years before the Vikings, Columbus, Cortez, and Pizarro? Needless to say, this is a most exciting and important question, not only for anthropologists and historians but for many nonprofessionals as well. There have even been some speculations that Caucasoid people may have been the original inhabitants of the New World. Thus, for scientists and others interested in such historical questions, further studies on Kennewick Man are overwhelmingly important.

For a while it looked as though these investigations would never take place. In an effort to comply with the federal Native American Graves Protection and Reparation Act, the Army Corps of Engineers assumed control of the bones, placed them in a vault, and refused to allow any further examination of them by scientists. The Umatilla tribe, who live near the site of discovery, asked to have the bones returned to them, in which case the skeleton would be secretly buried and never be available for study. A group of anthropologists went to court to stop the Corps from complying with the tribe's request. The anthropologists claimed that the Umatillas were not in that part of the Northwest when Kennewick Man lived; hence, he could not be one of their ancestors. And of course his Caucasoid skeletal features led to the same conclusion. Thus, the available scientific evidence is that Kennewick Man was not a Umatillan or any other Native American.

In response to that hypothesis, a leader of the tribe, Armand Minthorn, stated this position:

> Our elders have taught us that once a body goes into the ground, it is meant to stay there until the end of time. . . . If this individual is truly over 9,000 years old, that only substantiates our belief that he is Native American. From our oral histories, we know that our people have been part of this land since the beginning of time. We do not believe that our people migrated here from another continent, as the scientists do. . . . Scientists believe that because the individual's head measurements do not match ours, he is not Native American. Our elders have told us that Indian people did not always look the way we look today. Some scientists say that if this individual is not studied further, we, as Indians, will be destroying evidence of our history. We already know our history. It is passed on to us through our elders and through our religious practices. (Preston 1997, 74)

As of early 2001, the matter remains unsettled. Nevertheless the two perspectives—the anthropologists' and the Native Americans'—provide a classic example of two polar points of view that I will analyze throughout this book. One point of view rests on the questions and methods of science. The other rests on cultural beliefs that have been passed down from generation to generation. One side seeks a solution to a problem of intense interest to scientists and historians; the other side does not recognize that there is a problem that needs a solution.

This dispute over the future of Kennewick Man represents a clash between two immiscible patterns of thought. Another example (reported in the *New York Times* on August 29, 1997) comes from Afghanistan, where the Taliban, a Muslim sect, were in control of much of the nation and were enforcing conformity to the Sharia, sacred Islamic law. According to this code, thieves must be punished by having their hands and feet cut off; couples caught in adulterous acts must be stoned to death; and if women do not cover themselves from head to foot, the young Taliban enforcers dealt them a severe flogging. The young zealots even beat women for wearing white socks or plastic sandals.

The head of the General Department for the Preservation of Virtue and Prevention of Vice, Alhaj Maulavi Qalamuddin, explained the Taliban point of view: "Some women want to show their feet and ankles. They are immoral women. They want to give a hint to the opposite sex." This must be controlled to "prevent impure thoughts in men"; "if we consider sex to be as dangerous as a loaded Kalashnikov rifle, it is because it is the source of all immorality." The rules of the Sharia relating to women are harsh in other respects, by late-twentieth-century standards in the West. Women are prohibited from working or obtaining an education or even receiving medical treatment, and after puberty they are almost entirely secluded in their homes.

Such behavior toward women is not acceptable in most nations today, including most nations where Islam is the predominant religion. In this type of conflict modern values concerning human rights, gender equality, and civil liberties clash with religious doctrines that have been handed down for thousands of years. Such doctrines are accepted as "true" by the culture that inherits them and are highly resistant to change. In most instances, there is no way to adjudicate a conflict between these two systems of belief with evidence acceptable to each side.

Many—probably most—of the problems between nations, as well as the problems among the people of a single nation, stem from taking different points of view toward the same problem. Anthropologists using the methods and data of science propose one course of action for dealing with Kennewick Man, whereas the Native Americans relying on their received traditions propose another. Granted, only the scientists offer the possibility of reaching a decision about the identity of the remains based on confirmable data rather than on faith, but pursuing this course requires that both sides accept the validity of the anthropologists' assumptions, methods, and evidence; such acceptance by the Umatilla tribe seems unlikely. Consequently, in keeping with American jurisprudence, a federal judge will decide which point of view will prevail. By contrast, science has nothing useful to say when—as in the case of the Afghan women—the clash is between human rights and divine law. That conflict is between two different belief systems, each accepted as a matter of faith or principle, with one being far more restrictive on the lives of women than the other.

CREATIONISM VERSUS EVOLUTION

The immiscible patterns of thought that concern us in this book are those of creationists and evolutionists. Christian fundamentalists accept without question that divine creation is the explanation for the diversity of life we see today—the many different species of plants, animals, fungi, and microorganisms that flourish around the globe. Their position is based on their reading of Genesis, with its familiar story of the creation week—six days during which God created all of nature. On the first day God created heaven and earth and light and darkness; on the second He made the firmament and divided the waters; on the third day He separated land from the seas and created the land plants; on the fourth day He created the sun, moon, and other celestial bodies; on the fifth day animals in the sea and birds in the air came into being; on the sixth day the land animals appeared, and God also created, in his own image, two human beings. Until recently the accepted date for creation, based on a tally of the generations ("begats") listed in the

King James version of the Bible, was 4004 B.C. Creationists assume that all creatures living today are the same as when they were created. That is, there has been no evolution.

The scientific version of these events is quite different. Tentative estimates place the origin of the universe in the neighborhood of 15 billion years ago. The sun—a second-generation star—and its orbiting planets formed about 4.5 billion years ago from the interstellar debris created by the explosion of massive first-generation stars. The Earth started out in a molten state but cooled enough to form a solid crust by about 3.8 billion years ago. The first evidence of life dates to about 3.5 billion years ago and consists of very simple cells without a nucleus (prokaryotes), much like bacteria living today. The oldest known cells with nuclei (eukaryotes) date to about 2.7 billion years ago. The earliest multicelled organisms discovered so far date to about 1.7 billion years ago. Fossils of the first members of the animal kingdom date to about 650 million years ago. At the beginning of the Cambrian period, about 570 million years ago, animal life became abundant and highly diversified; the fossil record from this period is much better known than that of earlier times. From that point until today, there has been an incredible evolution of life (both plants and animals), with different species appearing, flourishing, and then becoming extinct or evolving into still other species.

The evolutionists' and the creationists' accounts for the origins and diversity of life could hardly be more incompatible. Strict creationists base their account on faith—a belief that Genesis was divinely inspired and provides the only true explanation for the origins of the universe, living creatures, and the many variations in organisms that we observe today and find in the fossil record. The creationists' pattern of thought begins with the answer and then seeks to explain the world in terms of that answer. Scientists work in the opposite direction. They study the universe and its earthly inhabitants and on the basis of observations and experiments propose a rational account for the past and the present. Scientists as a group, including those who adhere to religious beliefs, do not understand why any person familiar with the data would reject an explanation based on confirmable knowledge and accept instead a supernatural concept based on faith alone. Similarly, deeply committed fundamentalists wonder why anyone would reject the Word of God in favor of what a bunch of scientists has to say.

The long strands of human history have seen many conflicts between science and religion, and sadly, they have often turned violent and bloody. It is fascinating to consider how individuals come to hold such conflicting explanations for the same phenomena and why they hold them so tenaciously. We know that parents and society are remarkably efficient in transmitting patterns of thought and behavior to successive generations. This cultural inheritance sets up rules for behavior that help individuals get along within their group. It also provides each generation with an avenue for learning new things, and it creates order within the society. But beyond these practical advantages, a culture's unique belief systems may retain their enormous power from generation to generation in large part because they supply impressionable young people with answers to many questions they quite naturally ask, such as "Where did I come from?" "Who made me?" "Who will take care of me?" "Why do people die?" "What happens to me after I die?"

Such inquisitiveness seems to be part of our human inheritance. For hundreds of thousands of years, early human populations were illiterate and encapsulated in small tribes whose very survival was regularly endangered. Human beings lived a marginal existence like all other animals, dependent on the ability of the environment to sustain them and on their skills and knowledge of how to obtain food, water, and shelter. But within every society, some individuals must have exhibited a much stronger desire than their peers not just to survive from moment to moment but to understand themselves and the world around them. They asked questions and sought answers. According to anthropologists who have studied hunter-gatherer cultures, the explanations inquisitive tribe members come up with usually include both natural and supernatural elements. Natural things and processes are those that can be observed: wind, rain, birth, death, animals, plants, fire, night, day, and the seasons. But for each of these observable entities, a supernatural element of some kind usually figures in the explanation as well. For example, although death is now accepted by most people in Western societies as due to natural causes such as disease, accidents, or the ravages of age, people in some parts of the world still believe that death results from the displeasure of a god or spirit or the effects of a curse such as the evil eye.

The tenacity with which people hold onto beliefs in the supernatural or paranormal has been the subject of much scientific investigation. The results are complex and unexpected. Two psychologists, Barry Singer and Victor A. Benassi (1981, 49–51), made an extensive study of occult beliefs in the United States and offer this summary:

> Far from being a "fad," preoccupation with the occult now forms a pervasive part of our culture. Garden-variety occultisms such as astrology and ESP [extrasensory perception] have swelled to historically unprecedented levels. . . . Belief in ESP, for instance, is consistently found to be moderate or strong in 80–90% of our population; . . . in one survey it ranked as our most popular supernatural belief, edging out belief in God in strength and prevalence.
>
> Experiments which have attempted to encourage disconfirmation of occult or illusory beliefs by motivating subjects to think through their judgments more carefully . . . have uniformly revealed an astonishing resistance to change of such beliefs. . . . [Such] stubbornness of illusory and occult beliefs is typical rather than exceptional.

By way of explaining the origin of beliefs in the paranormal, Singer and Benassi point out that if human beings do not understand the reason for a given event, they tend to invent one, and the kind of reason they invent will depend on the intellectual baggage they carry with them. A simple, rational, natural explanation might be preferred, but if that cannot be developed, the need to explain remains and supernatural causes may be invoked. Thus, according to these investigators, one factor in the great increase of interest in the occult in the United States is inadequate science education in the schools. Over half of the students they tested did not know, for example, that the level of water in a partially filled glass remains parallel to the Earth's surface as the glass is tipped.

Singer and Benassi suggest that the "iffiness" of science is a second factor that weighs against the acceptance of scientific explanations over supernatural ones. Scientists tend to resist claiming that their statements are true by any absolute measure, stating only that they represent the best available explanation based on existing data. As Thomson (1997, 219) expresses it, the facts of science are "temporary way-stations on the long path to the refinement of knowledge." For many people, this tentativeness is a pale substitute for the finality and certainty of supernatural explanations.

A third factor that has increased interest in the occult is the media, which report stories about UFOs (unidentified flying objects), psychic healing, people who claim that the dead speak through them, and ghosts. Splendid examples of articles of this kind can usually be found in publications stacked at the checkout counters of supermarkets. The media rarely provide scientific analyses of these reports. Added to this "news from the other side" are highly entertaining motion pictures and television dramas that contain supernatural and superhuman feats. And of course even our own dreams can be quite extraordinary and entertaining, carrying us into compelling worlds where the constraints of waking reality no longer apply.

But Singer and Benassi found that the biggest factor by far in people's acceptance of the occult is organized religion. Most people profess a belief in some religion, and essentially all religions accept miracles as a given. This institutionalized belief in miracles, according to Singer and Benassi, has a spillover effect into other realms of life and accounts in large part for people's acceptance of paranormal events ranging from visits by angels to alien abductions.

Just how rigid occult beliefs can be is documented in an interesting experiment done with students in a psychology course at Concordia University in Montreal (Gray 1984). The course was called "The Science and Pseudoscience of Paranormal Phenomena." Students were given a test at the beginning of the class to see whether they believed in ESP, ghosts, or miracles (see table 1). At the end of the semester they were asked the same questions to see what effect studying these phenomena would have on their beliefs. Then, to get an estimate of the stability of any change in beliefs, the students were asked the same questions a year later. Occult beliefs declined by the end of the course but usually by trivial percentages. After one year, belief in the paranormal had moved back to levels close to those before the course was taken. In some cases it was slightly higher. This one-semester course designed to reduce students' belief in the paranormal was not a success. Lawson and Weser (1990, 589) report similar findings in another study and note that "the less skilled reasoners were more likely to initially hold the nonscientific beliefs and were less likely to change those beliefs during instruction. It was also discovered that less skilled reasoners were less likely to be strongly committed to the scientific beliefs."

Should we conclude that a willingness to accept supernatural explanations over scientific ones has been hardwired into the human brain—that it is "human nature" to seek meaning beyond the confines of the natural world? Perhaps, but an alternative—and to my mind more probable—hypothesis is that one's patterns of thought and belief are the result of influences very early in life associated with family, church, friends, community, books, other media, and the schools. Children almost always develop the habits and beliefs of the family and culture to which they belong; indeed, the fidelity of cultural inheritance often seems as strong as the fidelity of genetic inheritance. Most children in the United States grow up in communities where they see

TABLE I.
College Students' Belief in the Paranormal

Phenomenon	Students Who Believed in the Phenomenon		
	BEFORE THE COURSE	AFTER THE COURSE	A YEAR LATER
Extrasensory perception (ESP)	85	54	68
Unidentified flying object (UFO)	69	51	46
Astrology	55	43	61
Ghosts	43	34	45
Psychic healing	49	35	43
Miracles	43	46	54
Reincarnation/Life after death	69	55	57

many churches but encounter few or no institutions that extol the methods of science. Many television channels they watch feature evangelists expounding on miracles, heaven, and eternal damnation but show few scientists explaining science as a way of knowing. Most people are exposed to stories of creation from childhood, but few of them ever hear scientists' account of evolution.

NATURAL VERSUS SUPERNATURAL EXPLANATIONS IN WESTERN CULTURE

Natural and supernatural explanations for the way the world works have ebbed and flowed throughout human history, and both are still with us today. It is tempting to imagine a slow progress through time toward rational thought and naturalistic explanations and away from superstition, cults, and magic, but this is not the case. Shortly after World War II, the remarkable advance of science and technology elevated respect for the procedures and discoveries of science. But by the late 1960s, many people had become disenchanted with scientific perspectives and skeptical of the ability of scientists to do more good than harm.

Conversely, attempts to exclude the supernatural from explanations of natural events have been an important part of Western civilization from its beginning. As far back as the sixth century B.C., a few Greek philosophers who lived in Ionia on the coast of what is now western Turkey attempted to understand natural processes in terms of natural causes, a firmly rooted characteristic of science to this day. Only fragments remain of what they wrote, and most of the information we have about them comes from the surviving manuscripts of Greek authors who lived several centuries later. But from what scholars can tell, those Ionian thinkers were the earliest scientifically minded philosophers in the West.

The philosophy and science of the classical world of Greece and later of Rome continued this tradition. Aristotle in the fourth century B.C. codified the science of his time; and although later observations showed him to be incorrect in many instances, he sought natural explanations for natural phenomena. Throughout the centuries when Rome ruled the Mediterranean world, Western civilization saw a slow improvement in living conditions, technology, and general knowledge, all based to a great extent on a scientific approach to problem-solving. Those trends began to reverse themselves, however, in the fourth century A.D., with the passing of the great Roman emperors. By 476, after a long decline, the Roman Empire in the West fell and its political center moved east to Constantinople (modern Istanbul). The emperor Constantine was a Christian, and the Christian Church became increasingly powerful throughout the Middle East and the Mediterranean. The West had entered the Age of Faith. During the millennium from the fall of Rome to the fall of Constantinople in 1453, intellectual activity centered on the Church and was devoted to Christian practice and doctrine. Individuals who might have been the scientists, political leaders, artists, teachers, and philosophers in a more secular society were almost all occupied in the service of the Church. The literacy and education of the time were centered in the monasteries. Science and technology progressed little, and much of what had been generally known in classical times was forgotten.

The story was quite different farther east, in China, India, and Arabia, where learning flourished during this period. Islamic scholars made great contributions to mathematics and astronomy—a legacy still with us in the Arabic names of the major stars and in Arabic numerals. Eastern medicine was also superior to that in the West. Arabic scholars were greatly interested in what Aristotle, Plato, Galen, and other Greeks had written, and they translated the Greek manuscripts available to them into their own language. Later these works were translated from Arabic to Latin and so became available to scholars in the West. The quality of these classical works was so superior to any secular writings of the time that they eventually joined the Bible as the basic texts of the medieval period.

As writings of Greek and Roman scientists became available, some Christian scholars saw the need to adjust them to Christian beliefs. That great doctor of the Church Saint Thomas Aquinas (1225–1274) tried, for example, to adjudicate between the Averroists, who, following Aristotle, held that faith and reason were two absolutely distinct ways of thinking, and the Augustinians, who, following Saint Augustine (354–430 A.D.), believed that faith always precedes reason. Aquinas suggested a compromise: the truths of faith complement the truths of reason. This puzzling solution had advantages: in the short term, it protected the Averroists from being classed as heretics, with possibly grisly consequences; and in the long term, it saved Aristotle's work from banishment and kept the cause of reason alive in the Age of Faith.

By around 1400 A.D., another change in the intellectual climate began to be felt, leading to the period historians have traditionally called the Renaissance. This "rebirth" started in Italy in the late fourteenth century and then spread throughout Western Europe. Slowly but inexorably, humanistic themes replaced religious subjects in literature and the arts. Henry the VIII in England and Martin Luther and John Calvin on the Continent lessened the iron grip that the Catholic Church had exercised over those seeking new knowledge. Science and secular philosophy became acceptable pursuits once again. The year 1543 saw the greatest breakthrough in the cause of science, as three important intellectual works were published: the writings of the Greek physicist and mathematician of the third century B.C., Archimedes, which provided a basis for the later work of Galileo and Newton; Andreas Vesalius's text on human anatomy; and *De Revolutionibus Orbium Coelestrium* by Nicolaus Copernicus, a Polish astronomer and clergyman.

Both Vesalius and Copernicus proposed ideas that flew in the face of beliefs the Catholic Church had upheld almost without question for a millennium. These beliefs were based on the writings of Galen (b. 129 B.C. in Greece) on human anatomy and of Ptolemy (fl. 127 A.D. in Alexandria) on astronomy. When Michael Servetus, a Spanish physician, theologian, and scholar, made observations suggesting that Galen was not always correct in his anatomical descriptions, the Church had Servetus arrested, investigated, and in 1553, burned at the stake with a copy of one of his offending publications hung around his neck. When Giordano Bruno, a Dominican monk, rejected the Church-approved system of Ptolemy in which the heavens circled the Earth and accepted the system of Copernicus in which the Earth circled the sun, he too was burned alive, in 1600. Being burned alive is one of the most painful ways to die and, hence, appropriate for those who challenged authority in those times.

Still, a revolution in science had burst on the scene with a vigor the Church could not stamp out. By the time the seventeenth century came to a close, William Harvey (1578–1657) had described the circulation of the blood, Sir Francis Bacon (1561–1626) had written great treatises on the nature of science, and three pioneer astronomers and physicists—Galileo Galilei (1564–1642), Johannes Kepler (1571–1630), and Sir Isaac Newton (1642–1727)—had laid the basis for heavenly cosmology and earthly physics. The work of these scientists of the six-teenth and seventeenth centuries was for the most part grounded on careful observations and experimentation rather than on preconceived notions of how the world ought to work. Galileo, Kepler, and Newton discovered that complex phenomena, such as the motions of celestial and earthbound bodies as well as gravitation, obeyed precise rules that could be expressed mathematically. This was the beginning of serious attempts to reduce all natural processes to scientific—and ultimately mathematical—statements. Conclusions were based increasingly on natural processes and less and less on metaphysical concepts.

Suddenly all problems relating to nature seemed to be approachable by the unfettered human mind, and a sense of freshness and freedom swept over Western intellectuals. In 1664 Henry Power, Dr. of Physick, described how it felt to be part of this great intellectual revolution: "This is an age wherein (me-thinks) Philosophy comes in with the Spring-tide.... Me-thinks I see how all the old Rubbish must be thrown away, and the rotten Buildings be overthrown, and carried away with so powerful an Inundation" (192). Heady stuff. And he was not burned at the stake, mainly because he was an Englishman, and the Anglican Church was far less powerful in Britain than the Catholic Church was on the Continent. Times were changing, as Henry Power realized.

Science was stimulated by other developments as well. Western Europeans were voyaging around the globe, encountering new cul-tures and discovering hundreds of new species of plants and animals. The Royal Society of London, founded in 1662, became a center for scholars of Western Europe who were seeking a deeper knowledge of the natural world. Its publications served as a vehicle for making sci-entific knowledge available to all who might be interested. Universities throughout Europe were increasing in number and quality, and there was a slow but deliberate change in attitude about what was appro-priate intellectual activity.

RATIONAL VERSUS ROMANTIC THINKING

The triumph of rationalism in astronomy, mechanics, and anatomy spilled over into other human endeavors to such a degree that by the eighteenth century, Western civilization had entered a period known as the Age of Reason or the Enlightenment. The newly freed human mind no longer looked to the Bible or to the authority of ancient scholars for answers. Instead, it sought to study the natural and human worlds through the powers of logic and observation. Above all else, supernatural explanations were to be rejected. It became widely accepted among intellectuals that the universe operated according to natural laws, which could be discovered through scientific procedures and expressed mathematically. The knowledge gained could lead to great improvements in the human condition.

This is not to say that scholars of the Enlightenment uniformly rejected God or religion. Quite the contrary; they were usually pious Christians who believed that science was a useful tool for understanding God's work. According to the deistic worldview of the time, God could be known in two ways. One was to study "the Word," that is, the Bible, which was still accepted by the Church and nearly everybody else in the Judeo-Christian tradition as having emanated from God. The other was to study "the Work," that is, what He had wrought through creation. Scientists such as Galileo and Bacon were well aware of the many difficulties in interpreting what was said in the Bible, and they hoped that the "book" of nature might help to solve those puzzles. Interpreting nature, therefore, became a companion activity to the theologians interpreting the Bible. Together these pursuits would lead closer to truth—or so it was hoped.

Not surprisingly, the Age of Reason elicited a counter-reaction. The widespread attempt to "be scientific" in the eighteenth century was largely rejected by nineteenth-century intellectuals of the Romantic movement, who viewed the Enlightenment world as cold, heartless, and confining to the human spirit. In all this rationality, they asked, where was a place for inspiration, imagination, emotion, spiritual aspirations, self-expression, individual creativity, mystery, revelation, and tradition? Copernicus and his followers had demoted Earth from the hub of creation to a minor planet orbiting a minor star. The great eighteenth-century classifier of plants and animals, Carolus Linnaeus, had demoted humankind from the center of creation to a member of the Primate order, where a man or woman became just another ape. The Romantics accused science of finding no noble purpose for human life, overlooking the fact that finding purpose or creating meaning is not a proper assignment for science.

Human thought is far too complex to be divided strictly into these two discrete patterns—rational and romantic. Nevertheless, the pattern fairly typical among intellectuals of the Enlightment of the eighteenth and nineteenth centuries still characterizes much of our thinking in the modern world. During that time the pendulum swung noticeably toward the rational, critical, empirical, mechanistic, material, impersonal, skeptical, and reductionist mode. According to this ideal, unbiased observations and experiments produced data that others could confirm, and supernatural beliefs unsusceptible to scientific procedures were to be rejected. The world was a fit object of critical study, and not only was great progress possible in understanding it, but the products of science would inevitably increase the well-being of humanity. This philosophy—one might even call it a mission statement—is one that most scientists accept today.

But in some areas of human life, such as love, friendship, religion, the arts, and literature, a romantic way of thinking takes precedence over rational thought. Again, allowing for much fuzziness, one might characterize this kind of thinking as romantic, creative, emotional, humanistic, spiritual, personal, mystical, and holistic. But in no major undertaking does one mode of thought or action suffice. The design of an automobile that actually works is heavily weighted toward the operations of the rational mind; but if one hopes to sell the cars one manufactures, then style, beauty, and history must be given their due. The nurturing of a child requires both a romantic and a rational mind-set. Love and individual attention in a warm and supportive family situation are essential, but so is scientific information about nutrition and medicine for maintaining the child's health.

It is important to note also that intellectual debates between rationalists and romantics in the eighteenth and nineteenth centuries involved a mere fraction of society in a small corner of the world. Most human beings during that time lived, acted, and believed as human beings always had, and as most human beings do today. Moreover, we must not make the mistake of assuming that these two general patterns of thought first crystallized during the Enlightenment and Romantic periods. The Greek philosophers in Ionia were surely enlightened and they were just as surely capable of thinking in the romantic mode as well. Today, as then, the two points of view are different and often incompatible, but the consequences of that incompatibility are rarely severe. The human population is not divided into two hostile camps; everyone uses both patterns of thought, depending on what is being thought about.

The scientists I know make use of the romantic mode in at least 90 percent of their nonprofessional activities. The people they marry; the food they eat; the art, music, and literature they create or appreciate; the religious practices they follow; the recreation they enjoy; the pets they choose; and the clothes they wear are not selected on the basis of cold, calculating, impersonal data. And we can be thankful for that. On the other hand, few scientists I know—or anyone else, for that matter—would step off the curb into fast-moving traffic and expect a personal God to intervene and prevent their change from three dimensions to two.

Because all human beings are capable of these divergent thought patterns, one rational and the other romantic, there are no easy solutions for what to do with the bones of Kennewick Man or how to view the conditions of the women of Afghanistan under the control of the Taliban—or whether to teach creationism or evolution in the schools. There is no one acceptable answer to these social decisions, because different groups adhere, sometimes fiercely, to different answers.

TWO

Creation according to Genesis

Belief in a creation of some sort is not unique to the Judeo-Christian tradition. Most societies believe that the world and its living inhabitants have not existed forever, but rather that everything was created by some deity or supernatural force in the remote past. Many different scenarios for the mode of creation can be found; each one is usually restricted to a single culture and even to a period in history. A culture's explanation of creation becomes part of its sacred beliefs.

ORIGINS OF GENESIS

The Old Testament of the Christian tradition consists of the sacred scriptures of the Jewish people. The antecedents of Genesis, the first book of the Hebrew scriptures, date to the very dawn of recorded history and possibly to the early days of civilization in the ancient Near East, where the earliest cities seem to date to the few centuries before 3000 B.C. The Hebrews were the Semitic group from which both Judaism and Christianity arose. The Semitic people first entered history as tribes of nomads grazing their herds in the grasslands of Mesopotamia (now Iraq) and the Arabian peninsula. Five major groups are recognized: Akkadians, Canaanites, Phoenicians, Hebrews, and Arabs.

The Akkadians settled in Mesopotamia; the Canaanites, Phoenicians, plus the Hebrews eventually migrated west to the eastern shore of the Mediterranean. The Arabs continued to live in Arabia to the south. Much later, after the death of their prophet, Muhammad, in 632 A.D., the Arabs swept throughout the Mediterranean world as conquerers. They developed the Islamic religion with its holy book the Koran. Eventually most of the Semitic people abandoned the nomadic life of their ancestors and adopted agriculture, lived in cities, and developed complex political, social, and religious institutions.

The basic source for the early history of the Hebrew tribes is the Old Testament, plus a little information from archaeology and the records of neighboring people. For example, ancient Egyptian records, which are fairly extensive, make vague references to people who may have been the Hebrews. According to Old Testament sources and a modicum of other information, at least some of the Hebrew tribes seem to have been in ancient Sumer in southern Mesopotamia—possibly the site of the first civilization. About 1900 B.C. their leader was Abraham, according to ancient tradition. Later the tribes migrated westward to what is now Palestine and on to Egypt, possibly between 1700 and 1600 B.C. There, according to their traditions, they were enslaved. Sometime between 1300 to 1250 B.C. Moses led them back to Palestine, their Promised Land. Palestine was then known as Canaan, and it was occupied by the Canaanites, another group of Semites who were culturally more advanced than Moses' Hebrews, some even living in cities. Fertile land was scarce, and because the Canaanites were not inclined to abandon their territory to the newcomers, there was strife between the two groups for several decades. In fact, about all the Hebrews were able to conquer at first were the rather infertile hills to the east of the fertile lowlands that bounded the Mediterranean.

The period around 1200 B.C. was an exceptionally violent time in the eastern Mediterranean world for reasons not well understood. Some historians suggest that a mysterious "Sea People" traveled about widely, leaving destruction in their wake. The palaces of the Mycenaeans in what is now Greece were leveled; Troy fell; the Hittite Empire collapsed; and devastation extended from Greece through Turkey to the southern part of Palestine. Only Egypt survived relatively intact, having repulsed the Sea People in battles along the Nile delta. What had been an impressive Bronze Age, when the heroes described by Homer in the *Iliad* and the *Odyssey* lived, was replaced by a dark time that lasted several centuries.

One group of Sea People who entered Canaan during this period of turmoil were the Philistines. They proved to be far better warriors than the Canaanites or the Hebrews—a major reason being that their swords and spears were made of iron rather than bronze. The eastern Mediterranean world had embarked upon the Iron Age.

The Hebrew people, after losing some of the land they had conquered from the Canaanites, recognized the need for a more efficient political structure. Up to this time (about 1225–1025 B.C.) the Hebrew tribes had been loosely associated and led by charismatic religious and political leaders. Around 1025 B.C. the tribes began to consolidate, with Saul as their king. By the reign of David—next in line after Saul— the northern and southern tribes were united, and their armies successfully repulsed the Philistines. David and his son and successor, Solomon, were extravagant to the extreme and supported their projects with heavy taxation—a cause of considerable social unrest. So when Solomon died in 922 B.C., the united Hebrew kingdom split, with the ten northern tribes becoming the Kingdom of Israel and the two southern tribes, the Kingdom of Judah.

For the next eight centuries a series of invasions kept the Hebrews in turmoil. In 722 B.C. Assyria conquered the Kingdom of Israel and dispersed its inhabitants, who disappeared from history as organized communities. In 586 B.C., King Nebuchadnezzar of Babylon overwhelmed the Kingdom of Judah, destroyed Jerusalem, and took many prominent Jewish people back to Babylon—an episode known to history as the Babylonian captivity. Next came a period of Persian rule, from 539 to 332 B.C. Then Alexander the Great (356–323 B.C.) of Macedonia conquered most of the vast area from Greece to India and unified the Middle East both culturally and politically. After Alexander's death, Palestine was ruled by his generals, and much later it became part of the Roman Empire.

In 70 A.D. the Romans suppressed a Jewish rebellion and destroyed Jerusalem and the Temple. Many of the Hebrew people migrated to the various Roman provinces throughout the Mediterranean world— still another diaspora. What the Assyrians and Babylonians began, the Romans continued. Thus the Hebrew people were able to live free and united in their Promised Land for only relatively short periods— the few hundred years before 922 B.C. and less than a decade in the second century B.C. It is only in the twentieth century that the Jewish people have reestablished a nation of their own.

Scholars generally accept that the Hebrews began to develop their religious beliefs when they were seminomads. The first objects of their worship were probably the sun, the moon, springs, and mountains as well as mythological creatures and supernatural forces. In those early days, the gods or other objects of worship tended to be geographically restricted, with each cultural group worshipping its own gods. And so as they traveled the Hebrews would have encountered a bewildering variety of deities throughout the Middle East, from Atum and Horus in Egypt to Baal and El in Canaan.

The defining event in the Hebrews' religious development was the movement toward a single god, YHWH (or Yahweh when vowels are added). Moses is given considerable credit for this conversion from polytheism to monotheism. The name Yahweh has been variously translated as "He is," "He causes to be," or "I am who I am." Yahweh was in part a war god, but Jewish tradition credits Yahweh with transmitting to Moses many of the moral values and customs that influence our lives to this day—the Ten Commandments being foremost. The emphasis on moral values continued with the great Hebrew prophets Isaiah, Hosea, Amos, Micah, and of course Jesus.

There is scant information about the development of the religion of the Hebrews before the first millennium B.C. For a long time the sacred traditions were probably transmitted solely by word of mouth. Homer's *Iliad* and the *Odyssey* are examples of other stories transmitted orally long before they were produced in written form. An oral tradition implies the presence of special persons, such as priests and elders, who were the custodians and transmitters of sacred beliefs and religious practices. Oral transmission always carries a high risk of error, because the transmitter might forget some portions of very long stories and might alter others.

The word *bible* is derived from *byblos,* meaning a roll of papyrus, which together with leather, clay tablets, and stone were the writing materials in the ancient Mediterranean world. Byblos was also the name of a Phoenician city known for the production and export of papyrus. Single sheets were prepared by arranging layers of reeds from the papyrus plant at right angles to one another and then pressing and drying them. The result was a very durable and easily handled writing material. These flexible papyrus sheets could be attached to one another to form writing surfaces of various lengths, and the documents could be rolled and unrolled repeatedly. Pens were made from sharpened reeds, and ink was produced by mixing carbon with oil. The Bible, then, began as a collection of papyrus scrolls containing the sacred writings of the Hebrews.

Copying sacred texts onto papyrus scrolls was less prone to error then oral transmission, but complete accuracy was impossible. In the seventh century A.D., printing from wooden blocks was practiced in China and Korea. But in the West paper did not become available until the Middle Ages, and movable type was not invented until about 1450. Only with these technological advances could accurate and relatively permanent multiple copies of important documents such as the Bible be made.

Many of the sources that became the books of the Hebrew Bible, or Christian Old Testament, were produced as papyrus scrolls in the last millennium B.C. Three sources are of critical importance for the first portion of Genesis, where the stories of creation are told. They are known among scholars by the abbreviations J, P, and R.

The J source is the oldest and is thought to have been first produced in the tenth century B.C.—possibly between 950 and 800. *J* stands for *Jahweh,* the German spelling of *Yahweh* (many of the nineteenth-century scholars studying the origins of the Old Testament were German). This text was created shortly after the death of King Solomon and the separation of his realm into the kingdoms of Israel and Judah. *P* is for *priestly,* since it was the work of Hebrew priests who came later, possibly between 700 and 500 B.C. During those years the Hebrews were under the control of the Assyrians and the Babylonians. *R* is for *redactor,* referring to the editor or editors who in the fifth century, possibly between 450 and 400 B.C., combined J, P, and other sources to produce Genesis—probably in much the same form as we know it today. Two other major sources, E and D, are basic to the text of later books in the Old Testament but not to Genesis.

A word should be said about the first five books of the Bible: Genesis, Exodus, Leviticus, Numbers, and Deuteronomy. In the Hebrew tradition they are known also as the Law, the Law of Moses, the Pentateuch (meaning "five books"), or the Torah. They are of special significance to the Jewish people, as they are considered a direct communication from God through Moses to the Jews. In recent decades biblical scholars have questioned Moses' role in writing the Torah, in part because the last verses of Deuteronomy describe his death—and writing about one's own death is a difficult feat for any author to perform. Another work of major significance to Judaism is the collection of the opinions of rabbis relating mainly to moral issues and conduct. These opinions were orally transmitted until about 200 A.D., when they began to be preserved in written form as the Mishna; this was expanded to become the Talmud in the fifth century A.D.

The date when the Hebrew scriptures were first assembled to produce a single volume is unknown. Beginning in the sixth century A.D., successive generations of rabbis known as the Masoretes attempted to

compare the surviving documents and rid them of errors and differences to ensure the consistency of the texts. This approach was quite
different from that taken by the compiler of another sacred text, the
Koran, which consists of the revelations of God to Muhammad as
written down by his secretary. About 650–651 A.D. the Caliph Othman
assembled the secretary's notes and then took the precaution of de-
stroying all other notes and versions. Thus he did not have to deal
with a confusing and contradictory set of documents, as did the re-
dactor(s) for Genesis.

The oldest extant copy of the Masoretic text was transcribed in 950
A.D. by Aaron ben Asher. This was about two thousand years after the
first scrolls of the earliest Genesis stories may have been prepared, and
of course, they would have been copied numerous times in the interim.
Nevertheless, there is good evidence that the surviving Masoretic text
has a high level of accuracy: it is nearly identical with the Dead Sea
Scrolls. These scrolls date from roughly 250 B.C. to 135 A.D., and were
rediscovered from 1947 onward by Arabs in caves near the Dead Sea.
A copy of the Masoretic text, dated 1008 A.D., is the basis of today's
standard version of the Hebrew scriptures.

Far older is the Septuagint, a Greek translation produced in Alexandria, Egypt, beginning in the third century B.C. from Hebrew
originals that no longer exist. The Septuagint was prepared for Jews
living in Egypt who had lost their knowledge of Hebrew and needed
a version of their sacred traditions in a language they could read. That
language was Greek, the dominant language in the Near East after
the conquests of Alexander, a great promoter of Greek culture and
language. There is a wonderful story about the origin of the Septuagint, which means "70." The Egyptian king Ptolemy II (Philadelphus)
ordered the translation, which was supposedly completed by 70 or 72
scholars working independently for 72 days. The purpose of isolating
the scholars was to see if each of them was truly inspired by God: if
they were, all the translations should be identical. At the end of the
allotted time the 70 translations were examined and, lo and behold,

they were exactly the same. That convinced the king and all others that the translations must have been the inspired, accurate Word of God.

An important source of the Christian Bible is the Vulgate. This is a Latin translation made by Saint Jerome around 400 A.D., based on numerous variant Hebrew texts that are no longer available and also on the Septuagint. Another of his sources was the Hexapla, which had been prepared by the Christian scholar and philosopher Origen (185–254 A.D.). It consisted of six parallel columns, each with a different version of the sacred texts in Hebrew or Greek. Not knowing which column represented the true Word, Jerome had to make choices and change the text when he suspected that errors had been made by copyists. The usefulness of his work was subsequently much reduced by careless copying and other modifications. A somewhat improved Paris text of the Vulgate was printed in 1450–1452 by Gutenberg. It, too, changed over the centuries, but it remains the standard Bible of the Catholic Church.

Thus the Bibles of today have had a long and varied history. Generation after generation of scribes copied the sacred documents, introduced errors, corrected presumed mistakes, added new material, and made translations. The advent of printing reduced the variation, but it also meant that a few variants were arbitrarily selected as closer to the original meaning, while other versions were ignored. Today there are only a handful of standard translations in common use, and there are still significant differences among them.

The desire, of course, has always been to come as close as possible to what was originally said in the long-lost original Hebrew scrolls. This would not be an easy task even if an original scroll were available. Ancient Hebrew had almost become a ritual language before the time of Christ, and Aramaic was becoming the common language of the Jews. Moreover, Hebrew was originally written only with consonants, which could lead to serious problems in knowing what word was meant. Consider the problem of knowing the meaning of an English word written *lk*. It could be *like, lake, look, leek, or luke,* depending on which vowels are used. Macintosh (1972) provides this more complex example (slightly modified): "the cat was on the mat" would be written "thctwsnthmt" if the words were run together, as was the practice, and the vowels left out. Depending on the vowels inserted, the sentence could be read as "the coat was on the mat," "the cat was in the moat," or "the cut was in the meat." It was only about a thousand years ago that dots and dashes were added to the consonants of written Hebrew to indicate vowel sounds.

The problem of no vowels is actually not as serious as it might seem. Even when the text of Genesis had been written on scrolls, the oral tradition continued and the scriptures were read aloud. Thus the words written only with consonants were understood since they were also heard. So long as the spoken words were correct and the documents had not suffered through poor copying, accuracy would continue.

Nevertheless, the problem of knowing the meaning of many ancient Hebrew words continues to this day. Does an unintelligible word represent a copyist's error or is it a word for which the meaning has been lost? Scholars can sometimes resolve the problem by reference to other Semitic languages. For example, a word thought to mean only "to know" in Hebrew means both "to know" and "to be tamed" in Arabic. This suggests that Judges 16:9, which refers to Samson, could be translated "And his strength was not tamed" instead of "So his strength was not known." It should come as no surprise, therefore, that there are differences—often very important differences—between versions. For example, Isaiah 7:14 in the Septuagint reads "The virgin will conceive and bear a son," whereas in the Masoretic text *virgin* is replaced by *young woman.*

There are problems of translation even with the very first sentence in Genesis. The first phrase in Hebrew is *bereshith bara elohim.* Traditionally this has been translated "In the beginning God created." Some modern translations believe that a more accurate translation is

"When God began to create" (Alter 1996). This wording is the same as in a recent translation of the Torah, where the editors make this important point: "The implications of the new translation are clear. The Hebrew text tells us nothing about 'creation out of nothing' (creatio ex nihilo), or about the beginning of time" (Orlinsky 1969, 51). Thus, there was no beginning of the sort modern science identifies as a Big Bang. The second part of the first sentence is traditionally translated as "the heavens and earth." There is no authority for using the word *heaven,* since the Hebrew word *shamayim* means sky. So an accurate translation would refer to sky and earth. The editors suggest the intended meaning is probably the entire universe.

The amount of scholarship devoted to gaining a better understanding of the Bible is enormous, and this brief survey indicates only some of the difficulties encountered in trying to determine the correct text. In spite of all this work, uncertainty in understanding some of the ancient words and statements remains. The three major ancient versions of the Bible—the Septuagint, the Masoretic text, and the Vulgate—differ in detail, and it is usually not possible to determine which text is to be preferred. This has led the biblical scholar Keith R. Crim (1994, 23) to write:

> Because no original manuscripts of any of the books of the Bible
> are known to exist, and the ancient manuscripts that scholars use
> as the basis of translation differ widely at many points.... In spite
> of the wide agreement among scholars on textual questions, ex-
> amination of the notes in the margin of modern translations re-
> veals that scholars still differ as to which manuscript is the most
> accurate in a specific passage.

The opinions of biblical scholars are important in evaluating a fundamental tenet of creationists' belief, namely, the inerrancy of the Bible. Most likely no one in the last three thousand years has seen the original scrolls that were later combined as Genesis. So far as we know, those original manuscripts have joined the desert sands, and whatever truth they are thought to contain has to be accepted on faith alone; it is not confirmable on the basis of the available historical or scientific evidence. The only texts that creationists and all others have to rely on are the existing translations, whose history is checkered, to say the least. There are problems with the accuracy of copying, and there is uncertainty over the meaning of some of the ancient Hebrew words. Modern translations are the best that scholars have been able to provide, but the variations among them are appreciable. Still, since these texts are all that we have, if creationists and evolutionists want to compare their respective accounts of the diversity of life, it is to modern translations of Genesis that both groups must turn.

THE TWO CREATION ACCOUNTS IN GENESIS

Although the Bible is said to be the most widely read book in the Western world, few readers seem to notice that there are two very different accounts of creation in Genesis, one originating in the P scrolls and the other in the J scrolls. The P and J versions were combined to produce the first part of Genesis as we know it. Stephen Mitchell (1996) has recently provided a new, majestic, and beautifully written version of Genesis that identifies J, P, and other sources.

The story of how scholars came to the conclusion that Genesis has this complex composition is fascinating. It had been known for centuries that the Bible often presented two versions of the same event. This was a serious problem until the nineteenth century, when biblical scholars discovered that careful analysis of style can reveal significant differences between any two versions of a text. In the case of Genesis, this stylistic difference is most apparent in the original Hebrew, but subtle differences remain even in the translations available today. The P version account of creation consists of all of chapter 1 plus the first three verses and the first portion of verse 4 of chapter 2. The J version of creation begins with the second portion of verse 4 of chapter 2 and continues to the end of that chapter (see figure 1). One notable difference between P and J is in the name given to the deity.

GENESIS.

P VERSION	J VERSION

CHAPTER 1.

1 The creation of heaven and earth, 3 of the light, 6 of the Firmament: 9 the earth separated from the waters, 11 and made fruitful. 14 The creation of the sun, moon, and stars, 20 of fish and fowl, 24 of beasts and cattle. 26 Creation of man in the image of God; and his blessing. 29 The appointment of food.

In *a*the beginning *b*God created the heaven and the earth.

2 And the earth was *c*without form, and void; and darkness *was* upon the face of the deep. *d*And the Spirit of God moved upon the face of the waters.

3 ¶ *e*And God said, Let there be light: and there was light.

4 And God saw the light, that *it was* good: and God divided the ¹light from the darkness.

5 And God called the light *f*Day, and the darkness he called Night. ²And the evening and the morning were the first day.

6 ¶ And God said, *g*Let there be a ³firmament in the midst of the waters, and let it divide the waters from the waters.

7 And God made the firmament, and divided the waters which *were* under the firmament from the waters which *were* above the firmament: and it was so.

8 And God called the firmament Heaven. And the evening and the morning were the second day.

9 ¶ And God said, *h*Let the waters under the heaven be gathered to-

CHAPTER 2.

. . . in the day that the LORD God made the earth and the heavens,

5 And *c*every plant of the field before it was in the earth, and every herb of the field before it grew: for the *d*LORD GOD had not caused it to rain upon the earth, and *there was* not a man to till the ground.

6 But ²there went up a mist from the earth, and watered the whole face of the ground.

7 And the LORD GOD formed man ³*of* the dust of the ground, and breathed into his nostrils the breath of life; and man became a living soul.

8 ¶ And the LORD GOD planted a *e*garden eastward in Eden; and there he put the man whom he had formed.

9 And out of the ground made the LORD GOD to *f*grow every tree that is pleasant to the sight, and good for food; *g*the tree of life also in the midst of the garden, and the tree of knowledge of good and evil.

10 And *h*a river went out of Eden to water the garden; and from thence it was parted, and became into four heads.

11 The name of the first *is* Pison: that *is* it which compasseth the whole land of *i*Havilah, where *there is* gold;

12 And the gold of that land *is*

Figure 1.   The two accounts of creation according to Genesis in the King James version of the Bible, known as P and J accounts, shown side by side.

gether unto one place, and let the dry *land* appear: and it was so.

10 And God called the dry *land* Earth; and the gathering together of the waters called he Seas: and God saw that *it was* good.

11 And God said, *ⁱ*Let the earth bring forth ⁴grass, the herb yielding seed, *and* the fruit tree yielding fruit *ʲ*after his kind, whose seed *is* in itself, upon the earth: and it was so.

12 And the earth brought forth grass, *and* herb yielding seed after his kind, and the tree yielding fruit, whose seed *was* in itself, after his kind: and God saw that *it was* good.

13 And the evening and the morning were the third day.

14 ¶ And God said, *ᵏ*Let there be lights in the firmament of the heaven to divide ⁵the day from the night; and let them be for signs, and for seasons, and for days, and years:

15 And let them be for lights in the firmament of the heaven to give light upon the earth: and it was so.

16 And God made two great lights; the greater light ⁶to rule the day, and the lesser light to rule the night: *he made* the stars also.

17 And God *ⁱ*set them in the firmament of the heaven to give light upon the earth,

18 And to rule over the day and over the night, and to divide the light from the darkness: and God saw that *it was* good.

19 And the evening and the morning were the fourth day.

20 ¶ And God said, *ᵐ*Let the waters bring forth abundantly the ⁷moving creature that hath ⁸life, and ⁹fowl *that* may fly above the earth in the ¹⁰open firmament of heaven.

21 And God created great whales, and every living creature

good: there *is* bdellium and the onyx stone.

13 And the name of the second river *is* Gihon: the same *is* it that compasseth the whole land of ⁴Ethiopia.

14 And the name of the third river is *ʲ*Hiddekel: that *is* it which goeth ⁵toward the east of Assyria. And the fourth river *is* Euphrates.

15 And the LORD GOD took ⁶the man, and put him into the garden of Eden to dress it and to keep it.

16 And the LORD God commanded the man, saying, Of every tree of the garden ⁷thou mayest freely eat:

17 But of the tree of the knowledge of good and evil, thou shalt not eat of it: for in the day that thou eatest thereof ⁸thou shalt surely die.

18 ¶ And the LORD GOD said, *It is* not good that the man should be alone; I will make him an help ⁹meet for him.

19 And out of the ground the LORD GOD formed every beast of the field, and every fowl of the air; and *ᵏ*brought *them* unto Adam to see what he would call them: and whatsoever ¹⁰Adam called every living creature, that *was* the name thereof.

20 And Adam ¹¹gave names to all cattle, and to the fowl of the air, and to every beast of the field; but for Adam there was not found an help meet for him.

21 And the LORD GOD caused a *ⁱ*deep sleep to fall upon Adam, and he slept: and he took one of his ribs, and closed up the flesh instead thereof;

22 And the rib, which the LORD GOD had taken from man, ¹²made he a woman, and *ᵐ*brought her unto the man.

that moveth, which the waters brought forth abundantly, after their kind, and every winged fowl after his kind: and God saw that *it was* good.

22 And God blessed them, saying, Be fruitful, and multiply, and fill the waters in the seas, and let fowl multiply in the earth.

23 And the evening and the morning were the fifth day.

24 ¶ And God said, "Let the earth bring forth the living creature after his kind, cattle, and creeping thing, and beast of the earth after his kind: and it was so.

25 And God made the beast of the earth after his kind, and cattle after their kind, and every thing that creepeth upon the earth after his kind: and God saw that *it was* good.

26 ¶ And God said, *ᵒLet us make man ᵖin our image, after our likeness: and let them have ᵠdominion over the fish of the sea, and over the fowl of the air, and over the cattle, and over all the earth, and over every creeping thing that creepeth upon the earth.

27 So God created man in his *own* image, in the ʳimage of God created he him; ˢmale and female created he them.

28 And ᵗGod blessed them, and God said unto them, Be ᵘfruitful, and multiply, and replenish the earth, and subdue it: and have dominion over the fish of the sea, and over the fowl of the air, and over every living thing that ¹¹moveth upon the earth.

29 ¶ And God said, Behold, I have given you every herb ¹²bearing seed, which *is* upon the face of all the earth, and every tree, in the which *is* the fruit of a tree yielding seed; ᵛto you it shall be for meat.

23 And Adam said, This *is* now bone ⁿof my bones, and flesh of my flesh: she shall be called ¹³Woman, because she was taken out of ¹⁴Man.

24 Therefore ᵒshall a man leave his father and his mother, and shall cleave unto his wife: and ᵖthey shall be one flesh.

25 And they were both naked, the man and his wife, and were not ᵠashamed.

CHAPTER 3.

1 *The serpent deceiveth Eve,* 6 *Man's fall:* 9 *God arraigneth them.* 14 *The serpent cursed: his overthrow by the woman's seed.* 16 *Mankind's punishment; and loss of paradise.*

Now the serpent was more subtil ᵃthan any beast of the field which the LORD GOD had made. And he said unto the woman, ¹Yea, hath God said, Ye shall not eat of every tree of the garden?

2 And the woman said unto the serpent, We may eat of the fruit of the trees of the garden:

3 But ᵇof the fruit of the tree which *is* in the midst of the garden, God hath said, Ye shall not eat of it, neither shall ye touch it, lest ye die.

4 And the serpent said unto the woman, ᶜYe shall not surely die:

5 For God doth know that in the day ye eat thereof, then your eyes shall be opened, and ye shall be as gods, knowing good and evil.

6 And when the woman saw that the tree *was* good for food, and that it *was* ²pleasant to the eyes, and a tree to be desired to make *one* wise, she took of the fruit thereof, and did eat, and gave also unto her husband with her; ᵈand he did eat.

7 And the eyes of them both were opened, and they knew that they *were* naked; and they sewed fig leaves together, and made themselves ³aprons.

30 And *ᵂto every beast of the earth, and ˣto every fowl of the air, and to every thing that creepeth upon the earth, wherein *there is* ¹³life, *I have given* every green herb for meat: and it was so.

31 And ʸGod saw every thing that he had made, and, behold, *it was* very good. And the evening and the morning were the sixth day.

CHAPTER 2.

1 The first sabbath. 8 The garden of Eden. 17 The tree of knowledge forbidden. 19 The creatures named. 21 The making of woman, and institution of marriage.

Thus the heavens and the earth were finished, and all the host of them.

2 And ᵃon the seventh day God ended his work which he had made; and he rested on the seventh day from all his work which he had made.

3 And God ᵇblessed the seventh day, and sanctified it: because that in it he had rested from all his work which God ¹created and made.

4 ¶ These *are* the generations of the heavens and of the earth when they were created. . . .

8 And they heard the voice ᵉof the Lᴏʀᴅ Gᴏᴅ walking in the garden in the ⁴cool of the day: and Adam and his wife ᶠhid themselves from the presence of the Lᴏʀᴅ Gᴏᴅ amongst the trees of the garden.

9 And the Lᴏʀᴅ Gᴏᴅ called unto Adam, and said unto him, ᵍWhere *art* thou?

10 And he said, I heard thy voice in the garden, and ʰI was afraid, because I *was* naked; and I hid myself.

11 And he said, Who told thee that thou *wast* naked? Hast thou eaten of the tree, whereof I commanded thee that thou shouldest not eat?

12 And the man said, ⁱThe woman whom thou gavest *to be* with me, she gave me of the tree, and I did eat.

13 And the Lᴏʀᴅ Gᴏᴅ said unto the woman, What *is* this *that* thou hast done? And the woman said, The serpent beguiled me, and I did eat.

14 And the Lᴏʀᴅ Gᴏᴅ said unto the serpent, Because thou hast done this, thou *art* cursed above all cattle, and above every beast of the field; upon thy belly shalt thou go, and ʲdust shalt thou eat all the days of thy life:

15 And I will put ᵏenmity between thee and the woman, and between thy seed and her seed; it ˡshall bruise thy head, and thou shalt bruise his heel.

16 Unto the woman he said, I will greatly multiply thy sorrow and thy conception; in ᵐsorrow thou shalt bring forth children; and thy desire *shall be* ⁵to thy husband, and he shall rule over thee.

17 And unto Adam he said, Because thou hast hearkened unto the voice of thy wife, and hast eaten of

P VERSION	J VERSION
	the tree, of which I commanded thee, saying, Thou shalt not eat of it: cursed *is* the ground for thy sake; [n]in sorrow shalt thou eat *of* it all the days of thy life; 18 Thorns also and thistles shall it [6]bring forth to thee; and thou shalt eat the herb of the field; 19 In the sweat of thy face shalt thou eat bread, till thou return unto the ground; for out of it wast thou taken: for dust thou *art*, [o]and unto dust shalt thou return. 20 And Adam called his wife's name [7]Eve; because she was the mother of all living. 21 Unto Adam also and to his wife did the LORD GOD make coats of skins, and clothed them. 22 ¶ And the LORD GOD said, Behold, [p]the man is become as one of us, to know good and evil: and now, lest he put forth his hand, [q]and take also of the tree of life, and eat, and live for ever: 23 Therefore the LORD GOD sent him forth from the garden of Eden, to till the ground from whence he was taken. 24 So he drove out the man; and he placed at the [r]east of the garden of Eden [s]Cherubims, and a flaming sword which turned every way, [t]to keep the way of the tree of life.

The Hebrew word *Elohim* appears in the P version, while *Yahweh* appears in the J version. The translators of the King James Bible retained this difference by translating *Elohim* as "God" and *Yahweh* (which does not appear until the last sentence of verse 4 of chapter 2) as "Lord."

In the P version, the familiar account and the one most widely recognized by creationists, creation occurs over six days:

On the first day, when the earth was dark, wet, and formless, God (Elohim) created light and night and day.

On the second day God made a firmament (heaven or sky) to separate the waters above from the waters below.

On the third day, God separated land and water and created plants.

On the fourth day, God created the sun, moon, and stars.

On the fifth day, God created aquatic creatures and birds.

On the sixth day, God created other animals and man.

On the seventh day, God ceased all his work.

The much older J account of creation differs from the P account in that there is no reference to days. Some scholars interpret J to imply that all creation occurred instantaneously, not in six days as in the first account. A literal reading however, does imply the following sequence:

The earth was barren and without plant life.

Then the Lord God formed Adam from dust.

The Garden of Eden, with all plants, was formed.

The Lord God, noting that it was not good for man to be alone, formed wild animals and birds out of dust.

None of the above being a satisfactory partner for Adam, the Lord God formed woman from one of Adam's ribs.

Note that in the J version the Earth already exists when the Lord God forms Adam from dust, and there is no mention of the creation of light, water, sky, land, or celestial bodies. By implication, then, these must already have been in existence. Note also that in the sequence of creation man is formed first, then the plants, animals, birds, and woman; in the P version aquatic creatures and birds are formed first, then all other creatures and man.

How are we to interpret these different accounts of creation? If we take every statement in the Bible literally, one or the other account must be correct, but both cannot be so. This is not a new theological problem. The early Fathers of the Christian Church tried valiantly to solve the dilemma long before it was realized that Genesis was woven together from two different sources. Most early theologians accepted what was later recognized as the P account as *the* creation story, but others found J preferable. Finally, it was agreed that both accounts must be accepted, since the Bible in its entirety was the Word of God. Saint Augustine and many others encouraged this point of view. Augustine's position was that nothing is to be accepted save on the authority of Scripture, since greater is that authority than all the powers of the human mind. It was hoped that an understanding of the apparently incompatible accounts of creation would come in time. Awaiting that day, it was necessary to believe both that God had created the earth and all else in six days and that creation had been instantaneous, as the second account was interpreted.

Andrew Dickson White, the famous historian, diplomat, and first president of Cornell University, gave a fascinating account of the early attempts to resolve the dilemma of the two accounts of creation (1898, 1:6): "Serious difficulties were found in reconciling these two views, which to the natural mind seem absolutely contradictory; but by ingenious manipulation of texts, by dexterous play upon phrases, and by the abundant use of metaphysics to dissolve away facts, a reconciliation was effected, and men came at least to believe that they believed in a creation of the universe instantaneously and at the same time extended through six days."

The presence in the Bible of conflicting creation accounts and other narratives is interpreted by biblical scholars as examples of compromises between different groups within the Jewish population—if you can't agree on which version to accept, retain the stories of both groups. Strict creationists reject this political analysis concerning the creation text. But if creationists insist on the inerrancy of the Bible, they cannot use the data of Genesis to support their position that P alone is the correct description of creation, since a second, conflicting account is also present. One cannot take God at His word in the P account and ignore what He says in J.

NOAH'S FLOOD

Most creationists believe that the great flood described in Genesis, commonly referred to as the Noachian flood, was a historic event that was worldwide in its extent and effect. Both the J and the P versions speak of a devastating flood that destroyed all life, including human beings, except for a few individuals of each species that were taken into the Ark by Noah and his family. In contrast with the accounts of creation in the first two chapters of Genesis, where P and J are kept separate, they are merged in chapters 6–9, which describe the flood with only a few inconsistencies. Nevertheless, biblical scholars are in general agreement that the separate versions can be recovered.

Both the J and P accounts of the flood begin with God's discouragement at what has happened to His creation, especially to humankind, which has fallen into evil. God repents ever creating human beings (J) and decides to destroy them, along with most living creatures, in a terrible flood. Noah alone among humans meets with his approval. God instructs Noah to make an Ark so that he, his family, and a few mating pairs of all species can ride out the flood. The P account provides a brief description of the Ark, and it was huge: 300 cubits long, 50 cubits wide, and 30 cubits tall. A cubit is the distance from the elbow to the end of the longest finger. If a cubit is taken as 20 inches, the Ark would be about 500 feet long, 83 feet wide, and 50 feet tall. Noah was to take his extended family plus seven pairs of clean beasts, one pair of unclean beasts, and seven pairs of birds (J); or alternately one pair of all creatures (P). Everything else was to be destroyed. Food would be taken for the family and the beasts (P). Water, presumably, would not be a problem.

As soon as Noah and company were safely enclosed in the Ark, vast quantities of water poured upon the surface of the earth, from both above and below. It rained for 40 days and nights, according to J; according to P, the waters did not subside for 150 days and nights. The flood was so great that even the highest mountains were covered (J). Then the waters receded, and eventually dry land appeared. Noah's family opened the Ark door and along with all the creatures on the Ark, returned to the earth to be fruitful and multiply.

The creationists' belief that the flood described in Genesis actually occurred raises enormous problems, the major ones having been recognized for centuries. The Ark—a very big boat even by modern standards—was probably large enough to hold a few individuals from all the species of mammals and birds likely to have been known to the ancient inhabitants of the Near East, but it would have been totally inadequate for all of the many millions of species known to exist today. Supposedly all individuals not lucky enough to have obtained passage on the Ark were destroyed. The J account (chapter 6, verse 7) says that man, beast, creeping things, and birds are to be destroyed, whereas P (chapter 6, verse 13) says that all flesh is to be destroyed, and (in verse 17) that all flesh in which there is the breath of life is to be destroyed. In both cases only the few individuals of each species that entered the Ark were to survive.

The most direct reading of these plans of God would seem to mean that all life not on the Ark was to be destroyed—even fishes, for they have the breath of life. Some scholars have argued that there was no need to worry about the fishes—they could handle a flood—but the huge number of insects and other invertebrates in existence seem to have been overlooked as possible passengers. And neither the J nor P authors considers plants. As far as we are told, neither plants nor seeds were taken on the Ark; yet when the flood subsided and Noah sent birds out to see if there was dry land, the dove came back with a freshly picked olive leaf (J). Although God promised that He would destroy every living substance that He had made, apparently not quite all living substances had been destroyed.

Yet another problem for which there is no natural solution concerns the creatures found in other parts of the world. After all, the entire world, including the highest mountain peaks, was apparently covered by the flood. That implies that all species from Australia, North America, South America, those parts of Europe and Asia distant from the Near East, sub-Saharan Africa, and the many islands of the world's oceans would have to travel many thousands of miles over land and sea to reach the Ark. And they had to do it in a week—the amount of time God gave Noah to prepare for the flood (chapter 7, verse 4).

And finally, where would all that water, which would cover the highest mountains, come from, and where could it go when the rains stopped? Geologists have no idea.

OTHER ACCOUNTS OF ORIGINS

Every distinct tribe anthropologists have studied and every early civilization archaeologists have discovered seem to have developed a complex mythology to explain human life and nature. These represent attempts by prescientific and preliterate societies to explain observations and experiences that were beyond the realm of their understanding. How could one explain the sun, for example? Where did it come from? Why did it appear and disappear with such extraordinary regularity? And of course where did human beings come from? Since every individual has a birth, could it be that human beings as a whole had a birth, a creation? Many cultures have a story of creation, in contrast to a belief that human beings and the world have always existed. These explanations of creation invoke supernatural causes for events imagined to have occurred long ago. These explanations become part of the tribe's sacred traditions and are transmitted orally by the elders to the younger generations. A few examples of the creation accounts of ancient societies, mainly of the Middle East, are given here, and some of them show relations to the Genesis accounts of creation and the flood.

The Mesopotamian Accounts of Creation

Ancient Mesopotamia (now Iraq), where the earliest known civilizations began, was the Land of the Two Rivers, the rivers being the Tigris and Euphrates. Several different Mesopotamian creation myths survive, the best known of which is *Enuma Elish,* named for the opening words that mean "when on high." This account may date from around 2000 B.C., but the oldest known copies are from at least a thousand years later. The nineteenth-century British explorers Austin Henry Layard and George Smith and their fellow worker Hormuzd Rassam discovered the first of these at Nineveh, in the ruins of the library of the Assyrian king Ashurbanipal (669–627 B.C.). Although in some ways he was a brutal king, he was also a notable scholar with a splendid library—in this respect, an Assyrian Thomas Jefferson. The library contained not books but records of various sorts inscribed on clay tablets. These were prepared by inscribing the wedge-shaped grooves of cuneiform script into the still-wet soft clay of freshly prepared tablets. These tablets are almost indestructible, especially if baked—which happened when the king's palaces and library were burned.

Though unreadable when the tablets were discovered, cuneiform writing was later deciphered by Henry Rawlinson and others. These scholars found that the king had assembled a valuable series of texts, including many copies of very old Mesopotamian legends. Seven of these clay tablets contain almost the complete text of *Enuma Elish.* The story begins before creation with the presence of the divine parents, Tiamat the mother and Aspu the father, and their one son, Mummu.

As is so often the case in early myths, the gods and goddesses are also natural objects. Thus Tiamat was a saltwater ocean and Aspu a freshwater ocean. Mummu probably represented mist. These three waters made up the physical world. A period of begetting by the gods resulted in a sizable group of divinities; one of special note was Ea (also known as Enki and Nudimmud), Tiamat and Apsu's great-grandson.

Later some of the younger gods became rather boisterous and fun-loving, much to the annoyance of the original divine pair. Tiamat showed a considerable degree of motherly understanding, but Apsu was determined to destroy the lot. That decision was most disturbing, quite naturally, to the targeted gods. Ea, however, put a spell on Aspu and at length killed him and assumed his position. Ea and his wife, Damkina, bore Marduk, who was to become the greatest god in the Mesopotamian pantheon.

Tiamat remained inconsolable after the death of Aspu and together with some other disgruntled gods, Kingu among them, plotted revenge against Ea. When the plan for battle became known, Ea and the gods loyal to him grew terrified of Tiamat—who could also become a dragon—and her followers. Negotiations attempted by Ea came to naught, and finally he turned to Marduk for protection. Marduk agreed to help his father, with the proviso that he would become the chief god and have control over the whole universe. That request was granted, but tentatively, since the gods were not entirely sure of Marduk's powers. To test his powers, they placed a garment in their midst and asked Marduk to make it vanish. He did. As a double check they also asked him to restore it. Marduk could do that, too, so the gods were assured of the strength of their champion—and they still had their garment.

Both Marduk and his great-great-grandmother Tiamat prepared for their titanic struggle. Marduk armed himself with bow, arrow, and club. He filled his body with flames, caught the four winds in a net, and made seven other winds. He mounted a storm-chariot, was preceded by lightning, caused a flood, and went searching for Tiamat. Her fellow gods were terrified of Marduk, but Tiamat was not. She accepted his challenge for a duel. Marduk caught Tiamat in the net containing the winds and caused them to blow into her body, which expanded her greatly. Then he shot an arrow through her mouth and down into her heart. That did it. Later he split Tiamat in two, the upper half of her body forming the sky and the lower half the earth. He also made the celestial bodies. All of the rebel gods were captured and forced to work in support of the victors. Tiamat's main supporter, Kingu, was taken before Ea and his arteries were cut. Kingu's blood was then used to create humankind. Human beings were forced to assume responsibility for serving the gods food and other necessities. This was a major task, since by this time the Babylonian pantheon had grown geometrically.

Biblical scholars find much of interest in *Enuma Elish.* For example, both Genesis and *Enuma Elish* accept that there was a creation—the universe as a whole was not eternal. Creation in both accounts starts with a watery chaos, a detail found in other Middle Eastern myths as well, followed by separation of the water into sky and land. Light forms before the sun and other luminaries are created in both accounts. There are, of course, major differences, such as the presence of only one God in Genesis and many in *Enuma Elish.* The parallels and differences have led scholars to pose four hypotheses: Genesis may have been derived in part from *Enuma Elish; Enuma Elish* may have been derived in part from Genesis; both may stem from a single, more ancient source; or the two may have no relation, their parallels being coincidental. There is no firm evidence so far to indicate which of these alternatives is correct, although many scholars prefer the hypothesis that some of Genesis derives from *Enuma Elish.*

Another Babylonian myth—the *Epic of Gilgamesh,* which may date to the third millennium B.C.—is remarkably similar to the Genesis account of the flood. Gilgamesh was a mighty hero—a human, not a god. There is even the possibility that the final redaction of Genesis might have been done in the library of that Assyrian patron of ancient legends, King Ashurbanipal. The discovery of the first clay tablets of the epic was made by Layard and his associates. One short section describes a flood that probably took place in Sumeria, the southern part of Mesopotamia. That area is a flat plain not much above sea level at the north end of the Persian Gulf. Moderate floods were common, and severe ones were known, especially the flooding of the Tigris River.

According to the *Epic of Gilgamesh,* long after human beings had been created they became a noisy rabble. The warrior god Enlil became so irritated that he decided to destroy the lot—every human being. However, the god Ea spoke to one man, the son of Ubara-Tutu, who lived in the city of Shurrupak and told him to build a large boat. It was completed in seven days and was rectangular, measuring 120 cubits on each side. Gold, wild and tame beasts, seeds, and supplies needed for the duration of the flood were brought on board, along with the family and the necessary craftsmen. For six days and nights it rained. By the seventh day all was silent; human beings not on the boat had been turned to clay. The boat came to rest on Mount Nisir. A dove was let loose, but not finding a place to rest, she returned to the boat. A swallow was then let loose, but it too returned. Later a raven was released. She found food and did not return. The doors of the boat were then thrown open, and the living creatures disembarked. The gods came, and a sacrifice was made. All were pleased, except for Enlil, who had planned for all human beings to be destroyed. However, Ea was able to placate him.

The resemblances between the floods in the *Epic of Gilgamesh* and in Genesis are striking, but as with *Enuma Elish* and the Genesis account of creation, scholars do not completely agree about the relationship between the two. Many similar stories, both oral and written, circulated in the ancient Middle East, and the priests of different cultures had opportunities to become familiar with many of them. This was especially so after 586 B.C., when the Babylonians captured Jerusalem and took many of the leading Jewish citizens back to Babylon, where they would have learned of the Mesopotamian myths.

Creation Myths of Egypt

The archaeological information available today indicates that civilization began in Egypt almost as early as it did in Mesopotamia—and almost as abruptly. The First Dynasty began in Egypt about 3100 B.C., after a long predynastic period. Hieroglyphic writing had been in use for a few hundred years. Thanks to the dry climate, which tends to preserve artifacts and even papyrus scrolls, a great deal of information is available to us about Egyptian history. Compared with Mesopotamia, where empires rose and fell, the history of Egypt was relatively serene. The center of Egyptian civilization, the Nile Valley, was protected to a considerable degree by deserts on the east, west, and south and by the Mediterranean on the north, all of which would have presented an obstacle to large-scale invasion.

The pantheon of Egyptian gods was immense and changed over the centuries of Egypt's ancient history. The total number of gods, great and small, has been estimated at about 3,000. The Nile Valley consisted almost entirely of agricultural villages. In addition there were few cities that were centers of religion and government. Each village had its special god that looked out for the interests of the inhabitants. The larger political divisions, the nomes, also had their own guardian deities. The two largest divisions, Upper and Lower Egypt, each had a special god with a distinctive crown. During those periods when Upper and Lower Egypt were united, a single god was shown with a composite crown. The name of the supreme god for the entire country varied, depending on the era: Atum, Ra (or Re), Path, or Khepri.

The surviving manuscripts and inscriptions deal mainly with the role of gods and goddesses in contemporary life. These roles were usually quite diverse and could vary with time and place. The deities were almost always associated with some animal or physical object.

For example, a major god Horus was depicted as a falcon. Atum, a mighty god who created not only himself but the world and its inhabitants as well, was also the sun god, and depictions often show him with a solar disk on his head. Osiris was at one time the moon god but later was associated with life in the afterworld; so, rather appropriately, he is often represented as a bearded mummy. Hathor had various roles, such as the goddess of the four quarters of the horizon. She was generally portrayed as having a human body and the head of a wild cow.

According to Egyptian accounts of creation that have come down to us from ancient times, at the very beginning all was water—just like in the creation accounts of Genesis and *Enuma Elish*. The goddess Nun ruled supreme, and she created the god Atum. In early dynastic times, Atum's mythic role changed and he became the original god, possibly self-created. While still alone, he mixed semen with dust and created his brother Shu, god of air and life, and his sister Tefnut, goddess of moisture. Another version has Atum creating his brother and sister from his spit. Shu and Tefnut produced Geb and Nut. When Geb and Nut were together, Nut on top and Geb on the bottom, father Shu separated them as sky and earth. They gave birth to the next generation, which included Seth and Nephthys as well as Osiris and Isis and their offspring, Horus. Continued reproduction by the gods eventually produced all human beings.

Egyptian gods endured for more than three thousand years—a considerably longer period than has elapsed from their eclipse to the present day. Egypt's political independence ended in 30 B.C., when it became a province of Rome. The ancient Egyptian gods slowly became less relevant as gods of other cultures took their place.

Creation Myths of Classical Greece

In the fifth century B.C. Greece's prominence was rising as Egypt's was declining. While many Greek philosophers were trying to understand the world in natural terms, most ordinary Greeks probably continued to accept the myths about their origins that had been handed down from ancient times—the sacred traditions of the tribe. One of the earliest legends is credited to the Pelasgians, whom the historian Herodotus believed to be the most ancient inhabitants of Greece. According to the Pelasgians, a goddess named Eurynome was the first deity. In the beginning there was nothing but a watery chaos. Eurynome divided the sky from the waters and danced on the waves. She captured the north wind and created from it a snake, Ophion. Eurynome and Ophion mated, and Eurynome changed into a dove and laid an egg. The egg was incubated by Ophion and when it hatched, there emerged from it animals, plants, earth, and the celestial bodies.

The mythology more commonly associated with the later Greeks passed from the oral to the written tradition with Homer and Hesiod, probably in the eighth century B.C. The creation myth in this tradition begins with the goddess Gaia, who emerged from chaos and produced the earth and oceans. With no biological helper she gave birth to Uranus, the god of heaven. Similar virgin births are not uncommon in early mythology. Like Nut and Geb, Gaia and her son-husband then started producing gods. Their first offspring were the Titans, a varied lot that included most importantly the earth goddess, Rhea, and the evil god Kronos. Relations between the Titans and their father, Uranus, were far from good. Gaia tended to side with her children, as a good mother naturally would. Kronos decided to attack Uranus and did so while Gaia and Uranus were making love. At that critical and exposed moment Kronos castrated Uranus. The divine penis and testicles were thrown into the ocean, and from them Aphrodite, the goddess of love, was born and came floating to shore on a seashell.

With Kronos as chief god, the earlier era of dominance by goddesses was over. He raped his sister, Rhea, and in time there were six children-gods. Many were eaten by Kronos, but later they escaped relatively unharmed from his body. Rhea was able to hide and thus save the last child, Zeus, who eventually overthrew Kronos and established himself as the ruling god on Mount Olympus. By this time a very large number of gods were endowed with all the good and evil qualities of human beings. They could be a rowdy lot, given to human passions, but with supernatural powers not available to ordinary mortals. The gods created various groups of human beings, but most were found wanting and were either hidden in the earth or sent to the Blessed Isles. The final creation was the Iron race, which was the progenitor of all humanity.

Andaman Islands Creation Myths

The last example of creation myths will be those of the former inhabitants of the Andaman Islands. These islands form an isolated 200-mile-long archipelago in the Bay of Bengal between India and Myanmar (formerly Burma). The native inhabitants, when first studied by anthropologists around the beginning of the twentieth century, had one of the least technologically developed cultures known, yet they had a complex mythology, so different from those of the Middle East. They used fire but apparently did not know how to kindle it. Fish were one of their most important foods, but they did not have fish hooks; instead they used bows and arrows to shoot fish in the shallow coastal waters surrounding their islands. They had little contact with other people before the late nineteenth century, by which time they had developed not only a violent hostility toward outsiders but also a taste for them. The outsiders were mainly shipwrecked sailors, who when caught were promptly slaughtered and eaten. In the middle of the nineteenth century, in an effort to stop such killing, the British occupied the Andaman Islands, exposed the natives to civilization, and by the early twentieth century had driven them to extinction.

The British anthropologist Alfred Reginald Radcliffe-Brown made a classic study of the Andaman Islanders in the years 1906–1908. He found that they had many very different legends about the origins of humankind, though there was no consensus among them. Here is one example:

> The first man was Jutpu [meaning "alone"]. He was born inside the joint of a big bamboo, just like a bird in an egg. The bamboo split and he came out. He was a little child. When it rained he made a small hut for himself and lived in it. He made little bows and arrows. As he grew bigger he made bigger huts, and bigger bows and arrows. One day he found a lump of quartz and scarified himself. Jutpu was lonely, living all by himself. He took some clay [kot] from the nest of the white ants and molded it into the shape of a woman. She became alive and became his wife. She was called Kot. They lived together at Teraut-buliu. Afterwards Jutpu made other people out of clay. These were the ancestors. Jutpu taught them how to make canoes and bows and arrows, and how to hunt and fish. His wife taught the women how to make baskets and nets and belts, and how to use clay for making patterns on the body. (192)

One of the variations on this origin legend accounted for the origin of animals: "Sir Prawn once got angry and threw fire at the people [the ancestors]. They all turned into birds and fishes. The birds flew into the jungle. The fishes jumped into the sea" (207).

The world of the Andaman Islanders was enormously complex. Most things and processes of nature had not only a physical being but a spirit as well. Whereas the belief systems of the major religions invested supernatural powers in one or more supernatural beings, the Andaman Islanders believed that natural objects also had supernatural powers. The legends that recorded their history involved a constant interplay of the forces of nature and of human beings. There was no elaborate pantheon of gods of the sort that characterized the religions of Mesopotamia, Egypt, or Greece. Some spirits might be more powerful or more important than others, but that was all.

After Radcliffe-Brown left the islands, he developed a working hypothesis to explain what he had observed: "Customs that seem at first sight either meaningless or ridiculous have been shown to fulfill most important functions in the social economy, and similarly I hope to prove that the tales that might seem merely the products of a somewhat childish fancy are very far indeed from being merely fanciful and are the means by which the Andamanese express and systemize their fundamental notions of life and nature and the sentiments attaching to those notions" (330).

MYTHS AND THE HUMAN HEART

In societies of the developed world—now characterized by electronic technology, an exploding understanding of nature, a wealth of literature in all fields, and the desire to base conclusions on rational data— the term *myth* has a pejorative ring. It has not always been so. In the long span of human history before literature and literacy were widespread, the myths of the tribe provided the only taste of what is still essential in our lives today: knowledge apart from the chase and the hearth. Myths explained the forces of nature as well as the complex interrelations of human beings with one another and with the rest of their living and nonliving environment.

In spite of their tremendous variety, myths share some interesting features. Most obvious is an element of the supernatural: myths invoke gods, demons, devils, or angels with powers not available to mere humankind. The gods are not constrained by the laws of nature. Their supernatural powers are in fact violations of what seems natural to us and what we are able to do. Yet if those powers are used for our welfare, we feel less lonely and less personally responsible for what is the fickleness of fate.

Creation myths are almost universal among the world's cultures. This is astonishing. It would be far easier, and perhaps more logical, to suppose that the world has always been in existence. But even the modern science of cosmology, which studies the origin and structure of the universe, proposes a creation story: that a Big Bang started the universe around 15 billion years ago. It is intriguing to ask why the notion of a creation even arose. We may never know the answer, but possibly the observable fact that some plants grow from tiny seeds and some animals develop from tiny eggs may have given our ancestors the idea that the world as a whole has an origin.

One commonality of creation myths is the movement from simplicity to complexity. In the words of David and Margaret Lemming (1994): "The basic creation story, then, is that of the process by which chaos becomes cosmos, no-thing becomes some-thing. In a real sense this is the only story we have to tell" (viii).

So what does the present survey of Genesis and of other creation stories reveal about the hypothesis of supernatural creation as described in Genesis? I believe it puts us in a position to examine this claim with the same rigor we would apply to any document from any culture. The conclusions I draw are as follows:

1.   There are two conflicting accounts of creation in Genesis, according to the J and P texts. One or the other may be correct, but both cannot be. There is no logical mechanism for making a choice, and there is no scientific evidence in support of either version.

2.   In ancient times, and to some extent today, different groups of people have developed unique stories of creation by a supernatural force in the remote past. There is no logical way of choosing one "correct" creation story from among the many accounts available.

3.   Stories of creation are part of the sacred traditions of individual groups, and each group accepts its own traditions on faith. Those accepting the P or J version of creation in Genesis, or any other story of creation, do so on the basis of belief, not on scientific evidence.

Theologian and biblical scholar Alan Richardson (1953, 34) addresses the evolution versus creation conflict this way:

When we understand the nature of the Genesis parables, we shall no longer suppose that there can be a conflict between "science" and Genesis. . . . Genesis is not a scientific account of *how* the world came into existence; if I want to learn how this happened, I must go not to Genesis but to science. It is misleading even to speak of Genesis as "pre-scientific," for Genesis is not concerned with scientific questions at all; it is not a collection of the guesses of primitive men at answers to scientific questions. It is dealing with matters beyond the scope of science; its theme is man's awareness of his existence in the presence of God, his dependence upon and responsibility toward God. This high theme is dealt with in the only satisfactory way in which the human mind can deal with it, that is, in religious symbols. We shall miss the whole point of Genesis if we either take the parables of Creation, the Fall, and the rest, literally or look upon them as primitive guesses at scientific or philosophical truth.

Most scientists would find Richardson's statement acceptable be-cause it distinguishes between what science is and what it is not. Sci-ence can say nothing about a god, gods, or any other supernatural phenomenon; the tools and methods of science have been honed to explore the natural world, not the supernatural realm even if there is such a thing. By the same token, Genesis brings no scientific evidence to bear on the history of life over time, the age of the Earth, or the diversity of life that we see today. It provides no acceptable explana-tions for the remarkable biochemical similarities in the cells of micro-organisms, plants, and animals, for the universality of the genetic code, for the peculiarities of the geographic distribution of species, for the basic structure of major groups of organisms and the variations shown in the individual species in the group, or for the cycles of life that maintain a rough biological and chemical equilibrium in the environ-ment.

The Bible has basic things to say about morality, spirituality, and religious practice, and for many people its wisdom provides structure and meaning to their personal lives. It is not, however, a textbook of science, and no serious scholar or theologian pretends that it is. There is a saying attributed to Galileo, and possibly Saint Augustine before him, that goes to the heart of the matter: "The Bible tells us how to go to Heaven, not how the heavens go."

THREE

Creationists Meet Mr. Darwin, 1859

During the nineteenth century, theologians and biblical scholars became engaged in serious study of the antecedents of the Bible as we know it, comparing the various ancient texts from which it evolved, seeking to understand the apparent contradictions, and attempting to identify those texts that might have been closest to the original. While these scholars were busy studying "the Word" to illuminate God's great act of creation, other pious Christians were studying "the Work" itself—the many species of living creatures—in order to worship God better through a deeper appreciation of His handiwork. These religious students of nature called themselves natural theologians.

Natural theology was of considerable interest to clergymen and others in England. It was based on the belief that since all life had been created by God, a detailed study of plants and especially animals could reveal some aspects of the mind of the Deity. Clergyman John Ray had argued just that point in 1691 in *The Wisdom of God Manifested in the Works of the Creation.* More than a century later another English clergyman, William Paley, archdeacon of Carlisle, published *Natural Theology: or Evidences of the Existence and Attributes of the Deity Collected from the Appearances of Nature* (1802). It included a great deal of data on the anatomy of organisms and their remarkable adaptations to the environment in which they lived. Paley emphasized that life was so complex that it could only be the result of a divine power— such a complex design must have a designer. How could one possibly explain a single organism or even one of its parts, such as the eye, without invoking an "intelligent Creator"? It was impossible to imagine that a human being or an eye could "just happen." As we will see later this notion has had a rebirth recently as I.D., or intelligent design.

In the 1830s still another clergyman, the Right Honourable and Reverend Francis Henry, earl of Bridgewater, left a large bequest for the publication of a series of volumes "On the Power, Wisdom, and Goodness of God, as manifested in the Creation; illustrating such work by all reasonable arguments, as for instance the variety and formation of God's creatures in the animal, vegetable, and mineral kingdoms; the effect of digestion, and thereby of conversion; the construction of the hand of man, and the infinite variety of other arguments; as also by discoveries ancient and modern, in arts, sciences, and the whole extent of literature." Eight of these Bridgewater treatises were written by well-known scientists and published. They dealt with astronomy, physics, meteorology, chemistry, mineralogy, geology, and biology. They organized the available scientific knowledge to show what God had accomplished and conversely, suggested that science made sense only in relation to the Deity.

The natural theologians thought of their work as parallel to that of the bookish theologians. Studying God's Work in order to better understand Him seemed a reasonable approach in the eighteenth and early nineteenth centuries, when serious questions were being raised about the Bible's accuracy. Some natural theologians even went so far as to argue that the statements of God, whether transmitted in oral or written form, were surely subject to greater errors at the hands of humans than God's visible handiwork in living nature. The Word might err, but the Work spoke the truth.

In one of the most ironic episodes in intellectual history, the vast body of information the natural theologians accumulated about the life histories of plants and animals constituted much of the database Charles Darwin would draw on to support his concept of evolution. The beautiful adaptations of plants and animals could not be denied; all that was required was to switch the explanatory hypothesis from divine will to natural causes. Whereas the natural theologians began with the answer—divine creation—and then used the data they had gathered from nature to support the answer they had already decided was true, Darwin began with the data of adaptation and followed them wherever they led.

Charles Robert Darwin (1809–1882) was the son of a well-to-do English physician. He began and then abandoned a career first in medicine and then in the Church. From childhood, Darwin's first love had been nature—he had a particular fondness for beetles—and for several years he sought an explanation for many of the puzzles that he, along with the natural theologians, had already observed in the world around him. In addition to the amazing adaptations of plants and animals to their environment, he noted the sequential change of fossils in the geological record, the seemingly hierarchical relationships of organisms from simple to complex, some surprising discoveries by embryologists about early development, and the similarity of species living geographically close to one another. To understand why these items puzzled Darwin and his contemporaries, and why evolution seemed to offer the key, we must travel backward in time momentarily to the Scientific Revolution.

PAVING THE WAY FOR EVOLUTION

Among the great scientific curiosities during the two centuries before Darwin's time were objects called "figur'd stones" or, as Darwin's contemporaries would call them, fossils. Some of the figures in these curious stones were very similar to the oysters, clams, and snails that were common along the sea coasts of the world, yet the stones were often found high up on mountains. Perhaps they had been deposited there during Noah's flood. But still, how could an oyster get inside a rock?

Robert Hooke (1635–1703), a prominent member of the Royal Society of London and best known to biologists as the discoverer of cells, was much interested in fossils. He and his contemporaries collected them and, by publishing and comparing their findings, developed the skills to distinguish what had originated from a living creature from what had not. They came to the conclusion that fossils were formed when an organism died in the sea or a lake, sank to the bottom, and was covered by silt. As the silt increased in depth, the pressure would slowly convert the buried layers of silt and the entombed creature to stone. During this process, the original material—bone, shell, or a portion of a plant—would itself turn to stone. That is, the original material would be replaced, with almost molecule-for-molecule exactness, by the chemicals in the silt. This process of petrification solved the puzzle of how an oyster could get inside a rock (see figure 2).

But other puzzles were more difficult to solve. While some fossils seemed to be identical to living species such as shellfish, for most fossils there was no known living representative. This presented the natural theologians with a serious problem, because the Bible, at Ecclesiastes 4:14, says: "I know that whatever God does endures for ever; nothing can be added to it, nor anything taken from it." This teaching was widely accepted to mean that there have been essentially no significant changes in any organisms since creation. Minor variations might occur in varieties and breeds, but race horses and draft horses were still horses and all varieties of roses were still roses. The fossil findings, however, suggested that some creatures had failed to endure forever—they had become extinct.

The natural theologians came up with a temporary answer. Though there was no living mollusk on the coast of England that resembled the fossil species found there, individuals of the species could still be living elsewhere, possibly off the coast of Africa, and had simply not been discovered yet. A few such examples had already been found.

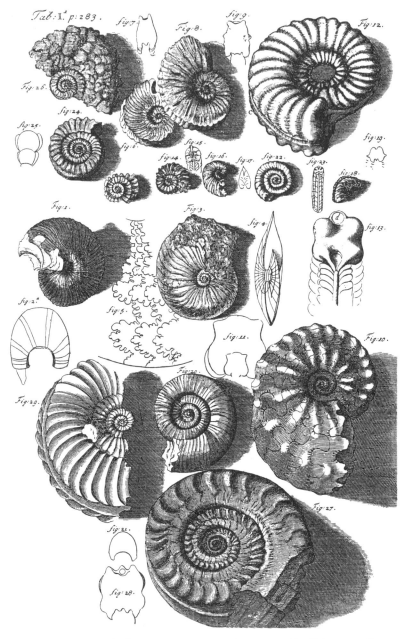

Figure 2. Fossils known to seventeenth-century English naturalist Robert Hooke. These are ammonites, a group of extinct mollusks related to still-living cephalopods such as nautilus.

But as more and more observations were made, it became highly probable that some fossils represented species that had become totally extinct. This was difficult to accept since it implied that God had made a mistake.

Two major types of rocks, volcanic and sedimentary, were recognized by geologists in Darwin's day. Volcanic rocks are formed by lava and ash from volcanic eruptions that became consolidated into rocks as they cooled. Volcanic rocks almost never contain fossils. In sedimentary, or stratified, rocks, on the other hand, the sediments are deposited in layers or strata, with the most recent stratum deposited on top. Fossils are found in sedimentary rocks, often in great abundance. By the nineteenth century it was clear that a given layer's position in sedimentary rocks provides valuable information about its age relative to the other layers—and hence about the age of any fossils it contains. Each layer of rock is older than the one above it and younger than the one below it, in the same way that the garbage at the bottom of a dumpsite is older than the garbage on top.

Most sedimentary rocks are formed at the bottom of oceans, in inland seas, or on continental shelves, where silt is washed down by rivers from higher elevations. We now know that at different times in the past much of the interior of the United States was covered by shallow inland seas. Each time the sea was present, sedimentary rocks would form at its bottom and preserve fossils of species prevalent during that time period. When the seas receded, the newly formed strata would be exposed. When subsequent changes in the Earth's crust elevated the strata in one area and exposed them to the surface, they became visible—and available for study by geologists.

Early in the nineteenth century geologists began to classify exposed rocks on the basis of their mineral composition and especially on the basis of the fossils they contained. The first major study of this sort was done in England by William Smith and published in 1815. He studied the strata in cliffs in numerous localities with the goal of arranging them in the order of their deposition. At one site he might recognize strata A, B, C, D, and E from bottom to top. At another site he might recognize F, G, H, I, and J. However, close study might show that F and G were identical to D and E, both in composition and in the fossils they contained. Thus he could conclude that the proper sequence, from oldest to youngest, was A, B, C, D = F, E = G, H, I, J.

Over the decades, this procedure produced the worldwide "geological column," an imaginary pile of strata that covers the span from the oldest discovered sedimentary rocks to those formed most recently. The height of this column—that is, the thickness of all known strata added together—is estimated to be about 60 miles (over 100 kilometers). This does not mean that one can start digging at any place on the Earth and go through 60 miles of strata. This impressive height is based on adding up all the different strata in the various places where they occur. How long did it take for all these strata to form? No one in the early nineteenth century knew, but geologists realized that it must have been a very long time, because silt washes into inland seas very slowly.

Continued studies established that each major group of strata contains its own unique kinds of organisms. The famous French naturalist Georges Cuvier (1769–1832) interpreted these data as evidence that at various times in the past drastic catastrophes had destroyed all life, and subsequently there had been new creations of quite different species. While Cuvier's theory, called catastrophism, held on to the idea of a divine creator for every single species, it significantly modified the creation process outlined in Genesis. Instead of taking just one week, creation in Cuvier's theory stretched throughout the entire history of life. And instead of recognizing just one great catastrophic flood, Cuvier suggested that life-destroying catastrophes had occurred over and over again.

An alternative explanation to catastrophism was evolution—the gradual change of species into other species over time. Darwin was not the first person to think of evolution; the concept had been around for centuries. Even the classical Greeks had speculated along these lines but then abandoned the idea when no data could be offered in support. The new observations and speculations of the seventeenth and eighteenth centuries, however, slowly laid the path toward a workable evolutionary theory.

One concept that helped pave the way was the *scala naturae,* or scale of nature—the suggestion that all animals could be arranged in a linear series based on increasing complexity, with no appreciable gaps in the series—from amoeba to humans. Where gaps seemed to exist, there were presumably intermediate forms yet to be discovered. Thus the great apes seemed to link human beings with other mammals, seals and whales linked fishes with land-living vertebrates, and bats were considered intermediate between birds and mammals. The roots of this concept could be traced back through medieval times to the Greeks, and it was still widely accepted in Darwin's day.

A further observation that prepared the way for evolution was that species of animals and plants are not randomly different from one another but seem to fall into naturally hierarchical groups. Similar individuals can be classified as the same species, similar species can be included in the same genus, similar genera in the same family, similar families in the same order, similar orders in the same class, similar classes in the same phylum, and similar phyla in the same kingdom. The first systematic attempt to classify living nature in this manner was made by the Swedish naturalist Carolus Linnaeus in the eighteenth century. In time, both the scale of nature and hierarchical classification were understood in terms of evolution—similar groups, such as species within a genus, are alike because they descended from a common ancestor. At the next level in the hierarchy, all the species of a genus of birds and indeed all species of birds, are descended from a very ancient common ancestor.

The person who first tried to bring ideas about evolution together into a coherent theory was the Frenchman Jean Baptiste Lamarck (1744–1829), who in his *Philosophie zoologique* (1809) maintained that one species evolves into another species in order to better adapt to its environment. Observing fossils in France, Lamarck noted that one geological stratum might have an abundance of one species of mollusk with little variation. The next higher stratum might contain species that were similar, but none would be exactly like those in the lower stratum. As he studied progressively higher strata, Lamarck observed that species became steadily different over time, stratum by stratum. Since fossils in a lower stratum were known to be geologically older than those in a higher stratum, it stood to reason that though a fossil in a higher stratum could not be the ancestor of one in a lower stratum—descendants cannot live before ancestors—a species in the lower stratum just might be the ancestor of a species higher up in the column. Lamarck concluded that what he was seeing in the fossils of progressively higher strata was change in a lineage over time. This hypothesis was markedly different from Cuvier's view that as the species in one stratum became extinct, closely similar ones were created anew and preserved in the next higher stratum.

Lamarck postulated a changing environment as the mechanism for the evolutionary change he observed. Species evolved in order to adapt, he believed. His classic example was the giraffe's remarkably long neck. The ancestors of today's giraffes, he said, had short necks and grazed on grasses and low shrubs, as do most other herbivorous mammals. Lamarck suggested that some ancestors of modern giraffes attempted to exploit a new and abundant food source—the higher leaves of trees. To reach the leaves they had to stretch their necks, which gradually lengthened with so much stretching. Lamarck thought that traits that came about through repeated use could be passed along to offspring. Thus giraffes would inherit the long necks of their parents and then stretch their own necks even further; over many generations, giraffe necks would become longer and longer until they reached the length of giraffe necks we see today. Conversely, characteristics that were not used would eventually wither away, as happened to eyesight in moles and bats.

This hypothesis of evolutionary change through "the inheritance of acquired characters" (meaning "characteristics") was not widely accepted in the early nineteenth century, since it was contrary to the Bible and was based on too much speculation and too few data. Other people besides Lamarck, including Charles Darwin's own grandfather Erasmus Darwin (1731–1802) had suggested that evolution might occur, but no one had yet argued the case well enough to convince the scientific community. Thus, in the first half of the nineteenth century the dominant scientific position was that species are "fixed," that is, they do not evolve. Although questions about the accuracy of the Genesis account of creation were being asked by scientists as well as biblical scholars, and alternative scientific as well as theological interpretations were being offered, in Charles Darwin's day none of these theories was taken seriously enough to undermine the traditional Judeo-Christian teaching. Evolution was out of favor; divine creation was still in vogue.

Nevertheless, in 1858 when Darwin sat down to prepare his treatise on evolution for publication, rich veins of biological and geological data were just waiting to be mined in evolution's support. Collections of animals and plants were being assembled and preserved in museums and herbaria at a prodigious rate—Cuvier himself had put together an outstanding zoological collection, especially of bones. Thousands of new species had been carefully described by taxonomists and placed in the Linnean system of classification, primarily on the basis of structural similarity. The embryonic development of the major types of animals was beginning to be understood. The main areas yet to be well understood were ecology, behavior, physiology at both the cellular and organ level, the study of microorganisms, and especially inheritance. Ignorance of the mechanisms of inheritance—or as we would now say, genetics—was to prove a serious problem for Darwin.

The geological data Darwin would draw on to support his theory of evolution had been well summarized in Charles Lyell's *Principles of Geology* (1830–1833). Lyell, an Englishman, saw no geological evidence for the vast catastrophes that Cuvier had invoked to explain the sequence of unique fossil faunas in strata of different ages. Lyell believed that all the usual geological events—volcanic eruptions, earthquakes, erosion by wind and water, and the uplift of land—were adequate to explain former changes in the Earth's surface, just as they explain them today. Thus, Lyell was a uniformitarian—a supporter of gradual geological change—not a catastrophist like Cuvier. Darwin, convinced by the arguments of his friend and countryman Lyell, would also favor gradual change over time and would reject sudden leaps in evolution.

Beyond these advances in biology and geology, science in general was becoming better organized by the middle of the nineteenth century. Publications, the life blood of any science, were numerous and of high quality, and biology and geology were being taught in universities. The Western world was in prime shape for a paradigm shift.

THE ORIGIN OF SPECIES

Charles Darwin did not set out to prove evolution. As a young naturalist, he thought he knew, as did all his contemporaries, that species are fixed to the degree that one species does not change into another. He knew, of course, that domestic species could be selected to produce strikingly different varieties. Horses could be selectively bred for speed or for strength. Roses could be selectively bred to climb higher, be more beautiful, or smell sweeter. Dogs and even pigeons could be bred to exhibit all sorts of new shapes and behaviors. But wild species of plants and animals were thought to be uniform, despite minor variations in individuals, and to remain essentially unchanged for centuries. For example, the ancient Egyptians embalmed animals of many species. When these were examined closely in the nineteenth century, three thousand years later, they seemed to be identical with contemporary individuals. These data were especially important in buttressing the Genesis account, since it was assumed that the early Egyptians lived not long after creation itself.

During the 1830s, when the Bridgewater treatises and Lyell's *Principles of Geology* were being published, Darwin embarked on the formative event of his scientific life—a voyage around the world on the *H.M.S. Beagle,* with Captain Robert Fitzroy in command. The purpose of the voyage was to map coastlines, for reliable navigational charts, necessary to avoid shipwrecks, were unavailable for most coastlines at the time. The young Darwin went along as the ship's naturalist, a common role in those days. It was on this voyage that he made some very puzzling observations that suggested that species might not be "fixed" after all. Fortunately, he left a written record of what had set his thoughts along this new track.

While the *Beagle* was making accurate navigation charts of the South American coast—a slow process with the instruments then at hand—considerable time was available for Darwin to go on shore to observe and collect living and fossil species. During these forays he made two sorts of intriguing observations. One had to do with the armadillos living on the Argentine pampas and a fossil armadillo, the glyptodont. These two sorts of mammals had many features in common, and Darwin assumed they were related. Both the living armadillos and the fossil armadillo-like glyptodonts were found only in the New World, mainly in South America. Darwin's inquisitive mind was always seeking answers, and he might have thought, "Is it not surprising that two such similar forms should have been created in precisely the same part of the world? Could the extinct species perhaps be the progenitor of the living?"

The second class of observations had to do with geographic variations within a species, a phenomenon Darwin first noticed on the mainland of South America. He observed that individuals of what were clearly the same species might vary from locality to locality—and the greater the distance between the localities, the greater the differences. But the most dramatic example of geographic variation Darwin observed was in the Galapagos, a cluster of small islands off the west coast of South America. Most of the species he found there were new to science and restricted to the Galapagos, although they were similar to species on the mainland of South America, 600 miles to the east. Some species were restricted to one island, although a similar species might occur on an adjacent island. Darwin found it astonishing that these islands, so close that most of them were in sight of each other, formed of the same kinds of rocks, with a similar climate, would have recognizably different varieties. Another dramatic example of geographic variation on a small scale was the Galapagos tortoises on islands close to one another. Minor differences in structure were such that the local inhabitants could tell the island of origin of any individual tortoise.

The hypothesis that Darwin proposed to account for the armadillo-like fossils and the Galapagos finches and tortoises was evolution. Darwin revived this generally moribund notion by suggesting a plausible mechanism for how it could occur. His argument went somewhat as follows: First, the environment on Earth is fixed in size and resources; there is only so much land and sea where organisms can live, find food, and reproduce. Second, every species has the potential for increasing its population size far above a level that the fixed environment can support. A single oak tree in a forest can produce thousands of acorns every year, yet in a mature forest no new tree can reach maturity unless an old tree dies and leaves a space. These two points were well known but by themselves they did not add up to evolution; those few lucky acorns that became trees would be the same kinds of trees as before. Evolution requires change, not a repetition of the same kinds of individuals from generation to generation. Something more was needed to bring about change.

The missing factors were variation and natural selection. Darwin had observed that individuals of a species vary slightly, that some variants are better adapted for surviving and producing offspring in a given environment than others, and that variations can be transmitted to offspring. In Darwin's day there was little solid information about how inheritance works, but animal and plant breeders knew very well that parents transmitted "something" to their offspring that influenced the characteristics of the next generation. They also knew that not all offspring inherited the desirable characteristics to the same degree; even among siblings there was variation, and this difference gave the animal or plant breeder a basis for selecting "the pick of the litter" for his breeding stock. Variation and selection rounded out Darwin's theory by providing a mechanism for evolutionary change. The hypothesis in full says:

1.   There is neither enough space nor enough resources for all individuals that are born to survive.

2.   Some individuals are better able to survive and reproduce under the conditions of a given environment than others.

3.   Organisms that survive and reproduce pass along to some of their offspring traits that improve their chances of surviving and reproducing in turn.

4.   Over time, this differential survival and reproduction will change a species in ways that better adapt it to its local environment and make it different from closely related species in slightly different environments.

5.   Given enough time, the differences between the current generation and its ancestral generation become so great that we say a new "daughter" species has evolved.

6.   Similarly, over time the differences between one population and another nearby become so great that we say a new species has evolved.

In making his case for natural selection acting on individual variation, Darwin relied heavily on evidence from plant and animal breeders, whose practices were familiar to him. In the breeders' case, it was human beings, not nature, who chose the characteristics to be preserved by allowing only those individuals with at least the vestiges of the desired features to reproduce generation after generation; all the rest were culled. While the majority of naturalists did not believe that new species in the wild could result from a natural version of this artificial selection, Darwin suspected otherwise.

The *Beagle* returned to England in 1836, and Darwin proceeded to prepare several books based on his observations and collections, including *Zoology of the Voyage of the Beagle* and *Geological Observations.* Neither volume mentioned his developing ideas about evolution. Later he wrote in his autobiography that in 1837 "I opened my first notebook for facts in relation to the *Origin of Species,* about which I had long reflected, and never ceased working on for the next twenty years" (Barlow 1958, 83). In 1842 and 1844 he wrote drafts of his ideas and told a few close friends, including Lyell and the botanist J. D. Hooker, about the possibility of the evolution of new species. In 1856 Lyell suggested to Darwin that he finish his studies and publish them before someone else anticipated his conclusions. Darwin took that advice and began to prepare a manuscript that was to become *On the Origin of Species.* The following year he sent a long letter describing his conclusions to the American botanist Asa Gray of Harvard.

Nevertheless Darwin's hypothesis for evolution still remained known to very few people. Various reasons have been suggested for his reluctance to broadcast his ideas more widely. One had to do with his beloved wife, Emma, who would have been upset with a notion so at variance with her deeply held Christian beliefs. One of the Darwin daughters wrote of her mother, "In her youth religion must have largely filled her life, and there is evidence in the papers that she left that it distressed her in her early married life to know that my father did not share her faith" (Barlow 1958, 239). Emma was not the only person upset by the direction Darwin's thoughts were taking. Many educated people in England, scientists and nonscientists alike, who read the *Origin* when it was finally published found it sorely distressing, since it undermined one of the fundamental beliefs of Western culture. And indeed there is considerable evidence that Darwin's conclusions at first distressed him as well.

Another explanation for Darwin's reluctance to publish is that the concept of evolution was held in low repute; the versions of evolution suggested by Lamarck in France and later by Robert Chambers in Britain had been vehemently rejected by scientists in England. Because Darwin was beginning with a widely rejected notion, it was all the more obligatory that he make an exceedingly strong and well-documented argument. Consequently, a vast amount of reading and thought had to go into the *Origin*'s production—a time-consuming process. Adding to this burden was Darwin's poor health. The medical profession continues to speculate on the nature of his illness to this day; the diagnosis varies from a parasitic disease contracted in South America to psychological problems caused by a domineering father. After Darwin had returned from the *Beagle* voyage, he and Emma had moved from London to the country—to the village of Down— where Darwin had become almost a recluse by the time the *Origin* was published. His ill health may have made him reluctant to subject himself to the stress of intense criticism from his scientific peers and others in his social circle.

A final reason for Darwin's slowness was that during the twenty years he thought about evolution before he started the actual writing of the *Origin,* he wrote other major works based on his findings during the *Beagle* voyage (two on geology and four on barnacles), and he conducted many scientific experiments to test his various hypotheses. All in all, it was not a bad rate of production for a graying, sickly recluse.

Despite these impediments, Darwin carried through on Lyell's suggestion, and by 1858 the book manuscript was about half complete. Then a bombshell hit. In June of that year Darwin received in the mail a manuscript from Alfred Russel Wallace, an English naturalist who was then collecting in the Malay Archipelago. Unlike Darwin, Wallace came from a modest social background and for many years had made a living by collecting in the New World and the East Indies and selling his specimens to gentlemen collectors and museums. Darwin was thunderstruck with what he read in Wallace's manuscript, as he explained in a letter to Lyell dated June 18, 1858 (F. Darwin 1888, vol. 1, 116–17):

My dear Lyell.

—Some year or so ago you recommended me to read a paper by Wallace in the "Annals" which had interested you, and, as I was writing to him, I knew this would please him much, so I told him. He has today sent me the enclosed manuscript, and asked me to forward it to you. It seems to me well worth reading. Your words have come true with a vengeance—that I should be forestalled. You said this, when I explained to you here very briefly my views of "Natural Selection" depending on the struggle for existence: I never saw a more striking coincidence; if Wallace had my MS. sketch written out in 1842, he could not have made a better short abstract! Even his terms now stand as heads of my chapters. Please return me the MS., which he does not say he wishes me to publish, but I shall, of course, write and offer to send to any journal. So all my originality, whatever it may amount to, will be smashed, though my book, if it will ever have any value, will not be deteriorated; as all the labour consists in the application of the theory. I hope you will approve of Wallace's sketch, that I might tell him what you say.

My dear Lyell, your most truly,
C. Darwin.

Darwin feared he had been scooped. He had developed what was to become the most important theory ever formulated in biology, and before he had made his ideas known, Wallace had reached almost identical conclusions. In science, the rewards go to those who first publish an important discovery or a new theory. Darwin could go ahead and publish first, of course, but if he did, how was Wallace to be given credit for developing closely similar ideas? For Darwin this was a terrible dilemma. Moreover, he was in his usual poor health, and the timing of Wallace's letter proved more difficult for him than it might have to a healthier man. And to augment the anguish, his son Charles, only half a year old, died ten days after Wallace's letter arrived.

Darwin turned to his friends Lyell and Hooker for advice, and they suggested a solution, namely, that they would send both the Wallace and Darwin manuscripts to the Linnean Society of London for joint publication. Darwin agreed, and the two manuscripts were delivered to the society on July 1, 1858, and published shortly thereafter. Part of the letter of transmittal reads as follows (Darwin and Wallace 1958, 258):

> So highly did Mr. Darwin appreciate the value of the views therein set forth [in Wallace's essay], that he proposed, in a letter to Sir Charles Lyell, to obtain Mr. Wallace's consent to allow the Essay to be published as soon as possible. Of this step we highly approved, provided Mr. Darwin did not withhold from the public, as he was strongly inclined to do (in favour of Mr. Wallace), the memoir which he himself had written on the same subject, and which, as before stated, one of us had perused in 1844, and the contents of which we had both of us been privy to for many years. On representing this to Mr. Darwin, he gave us permission to make what use we thought proper of his memoir, etc.; and in adopting our present course, of presenting it to the Linnean Society, we have explained to him that we are not solely considering the relative claims to priority of himself and his friend, but the interests of science generally; for we feel it to be desirable that views founded on wide deduction from facts, and matured by years of reflection, should constitute at once a goal from which others may start, and that, while the scientific world is waiting for the appearance of Mr. Darwin's complete work, some of the leading results of his labours, as well as those of his able correspondent, should together be laid before the public.

We have the honour to be yours very obediently,
Charles Lyell, Jos. D. Hooker.

The presentation of the papers by Darwin and Wallace at the
Linnean Society and their subsequent publication seemed to arouse
little interest. Darwin knew of only one review, and its verdict was
"that all that was new in them was false, and what was true was
old" (Barlow 1958, 122). In any event, Darwin in great haste prepared a short version of his manuscript, which was published on
November 24, 1859, as *On the Origin of Species by Means of Natural
Selection, or the Preservation of Favoured Races in the Struggle for Life*.
In contrast with the lack of interest shown at the Linnean Society
meeting, the book must have been keenly anticipated because it sold
out on the day of publication; a second edition was ready the following month.

DARWIN'S RECEPTION

The *Origin* was indeed threatening to most people in the West, for
Darwin's arguments could be interpreted as implying a world without
either God or purpose. Two Harvard professors, Asa Gray and Louis
Agassiz, represent the two poles of reaction among Darwin's most
educated readers. Gray was the leading botanist of the time in the
United States. Agassiz, a geologist and zoologist, was born in Europe
and later came to the United States, where he had a sparkling career.
Gray's initial position was similar to that of most mid-century naturalists—orthodox in religious beliefs. But because he had corresponded
with Darwin before the publication of the *Origin,* he was generally
aware of the arguments that it would contain, and his review of the
book, published in 1860, was careful and fair. He was not fully convinced that Darwin was correct, but he felt that a powerful case had
been made for the possibility of evolution and that the matter should
be seriously considered (A. Gray 1860):

We are thus, at last, brought to the question; what should happen if the derivation of species [evolution of one species from another] were to be substantiated, either as a true physical theory, or as a sufficient hypothesis? What would come of it? The enquiry is a pertinent one just now. For, of those who agree with us in thinking that Darwin has not established his theory of derivation, many will admit with us that he has rendered a theory of derivation much less improbable than before; that such a theory chimes in with the established doctrines of physical science, and is not unlikely to be largely accepted long before it can be proved. Moreover, the various notions that prevail,—equally among the most and least religious,—as to the relations between natural agencies or phenomena and Efficient Cause, are seemingly more crude, obscure, and discordant than they need be. (180)

The work is a scientific one, rigidly restricted to its direct object; and by its science it must stand or fall. Its aim is, probably not to deny creative intervention in nature,—for the admission of the independent origination of certain types does away with all antecedent improbability of as much intervention as may be required,—but to maintain that Natural Selection in explaining the facts, explains also many classes of facts which thousand-fold repeated independent acts of creation do not explain, but leave more mysterious than ever. How far the author has succeeded, the scientific world will in due time be able to pronounce. (184)

Gray's review was widely praised. Darwin regarded it as the best that had been written by that time, even though Gray was far from endorsing the argument in its entirety. He emphasized the problems that Darwin had admitted and pointed out others himself. For example, he noted the critical lack of any real evidence of the origin and nature of genetic variation. Yet in spite of his own ambivalence, Gray insisted on a fair hearing for Darwin, and he became the most vigorous defender of Darwinism in America. He made the prophetic statement that was borne out by subsequent events: the hypothesis "is not unlikely to be largely accepted long before it can be proved."

Gray's principal opponent was his fellow Harvard professor Louis Agassiz, and the debates over Darwinism that were to ensue in America centered on these two individuals. Agassiz, like Gray, was conventional in his religious views. He also reviewed the *Origin* in 1860, but his approach was to demolish, not explain, the arguments.

> Had Mr. Darwin or his followers furnished a single fact to show that individuals change, in the course of time, in such a manner as to produce . . . species different from those known before, the state of the case might be different. But it stands recorded now as before, that the animals known to the ancients are still in existence, exhibiting to this day the characters they exhibited of old. The geological record, even with all its imperfections, exaggerated to distortion, tells now, what it has told from the beginning, that the supposed intermediate forms between the species of different geological periods are imaginary beings, called up merely in support of a fanciful theory. The origin of all the diversity among living beings remains a mystery as totally unexplained as if the book of Mr. Darwin had never been written, for no theory unsupported by fact, however plausible it may appear, can be admitted in science. . . . (144) It would be superfluous to discuss in detail the arguments by which Mr. Darwin attempts to explain the diversity among animals. Suffice it to say, that he has lost sight of the most striking of the features, and the one which pervades the whole, namely that there runs throughout Nature unmistakable evidence of thought, corresponding to the mental operations of our own mind, and therefore intelligible to us as thinking beings, and unaccountable on any other basis than that they owe their existence to the workings of intelligence; and no theory that overlooks this element can be true to nature. (146)

Both Agassiz and Gray came from conventional Christian backgrounds, but neither accepted the inerrancy of Genesis in explaining the origin and diversity of life. Agassiz, especially, had made geology one of his major research interests, and he was well aware that the fossil record showed different faunas in the different geological periods. This refuted the Christian belief that all species had been created at the same time and had remained unchanged to the present day. He was equally aware, however, of the absence of any fossil evidence that one kind of animal evolves into another. Fossils that were intermediate between major groups of organisms, as required by Darwin's theory, were nowhere to be found. Darwin himself was fully aware of these difficulties and made no effort to conceal the fact that they spoke against his hypothesis. For Agassiz, Darwin's theory failed to explain the variety of living things. His only suggestion, which had a powerful appeal for most individuals, was that the entirety of nature was due to the workings of intelligence. Thus he allied himself with Paley and the natural theologians.

The drama unfolded with even greater intensity in England. For most nonscientists the arguments offered in support of evolution often seemed irrelevant and difficult to understand: Mr. Darwin had entitled his book *On the Origin of Species,* but he admitted there was no direct evidence for even one species changing into another! Most scientists, including accomplished field naturalists, were equally unconvinced. The conventional view of the fixity of species accounted for the data quite satisfactorily, and the social pressures to believe that nature was created by Divine Will were hard to ignore. Consequently, powerful voices in science buttressed the popular reaction by proclaiming that Darwin's evidence would not survive careful scrutiny. Of the several reviews by important scientists that were published in the spring of 1860, most were negative.

In the many contentious debates about evolution that followed, many scientists along with most religious leaders and ordinary citizens stood on one side, while on the other side stood Darwin and a few stalwart supporters who believed that the concept of evolution provided a rational explanation for innumerable biological and geological facts and was worthy of serious study. The position of the supporters did not signify complete agreement that evolution by natural selection was true but only that it was a useful hypothesis to be tested.

The first notable public confrontation came on June 30, 1860—seven months after the *Origin*'s publication—at a meeting of the British Association for the Advancement of Science at Oxford University. The event must have been eagerly anticipated, since the original room scheduled for the debate proved far too small, and the speakers and audience were moved to a larger chamber. No account of what transpired was published, but the story that has come down to us, in exaggerated form no doubt, is as follows: The mood of the audience—scientists, members of the clergy, and laypersons—was distinctly anti-Darwin. Because health problems prevented Darwin from attending, it remained for Thomas Henry Huxley (later dubbed "Darwin's bulldog") to support Darwin's position. The principal speaker to critique Darwin's views was the bishop of Oxford, Samuel Wilberforce. He was a prominent clergyman with a silver tongue, which had earned for him the sobriquet "Soapy Sam." The fact that a prominent bishop rather than a scientist was chosen to refute Darwin's challenge shows that evolution threatened not just established scientific theories but the religious establishment as well.

Wilberforce knew very little about science, but the distinguished anatomist Richard Owen had coached him on the scientific arguments showing the improbability of evolution. The bishop did not fully understand Owen's arguments, it seems, and in any case he did not use them. But he did understand what evolution implied about human origins: that humankind was descended from ancient apelike creatures. Wilberforce's remarks to the very friendly audience were amusing and glib, and in his conclusion he turned to Huxley, who was sitting on the speakers' platform, and "begged to know, was it through his grandfather or his grandmother that he claimed his descent from a monkey?" The audience went wild. Asked by the chairman to respond, Huxley briefly outlined Darwin's views and then came in for the kill. After stating that he would not be ashamed to have a monkey for an ancestor, he turned to Wilberforce and added that he would, however, be ashamed to be connected with a man who used his great gifts to obscure the truth—implying that it was better to have descended from a monkey than from Soapy Sam. Huxley's remarks changed the mood of the audience dramatically.

Although nothing was solved by this superficial debate, many scientists realized that Darwin's position was novel and important enough to be considered carefully. T. H. Huxley became a dominant force in trying to help both scientists and ordinary citizens understand evolution. He was a gifted speaker and a first-rate biologist steeped in comparative anatomy, embryology, and general natural history. In 1860 Huxley gave a series of lectures, mainly related to evolution and other aspects of contemporary biology, to workingmen in London. These lectures were later published and reached an even larger audience. In 1863 he published *Evidence as to Man's Place in Nature,* in which he showed the close anatomical similarity of human beings with the great apes and suggested that this was sufficient evidence for all to be placed in the same family in the scheme of classification. Although Huxley was careful in drawing conclusions, a reader familiar with Darwin's work would have suspected that Huxley thought the resemblances of the great apes and human beings were a consequence of their inheritance from a common ancestor. As time passed, Huxley became less sure that natural selection was the driving force behind evolution, as Darwin had proposed, and he was far from being alone among scientists in holding such doubts.

More professional debates followed within the scientific community, where harsh and demanding critics evaluated Darwin's data. Scientists were no less troubled than laypeople by the prospect of replacing God with a natural theory, and many went to great lengths to reconcile Darwin's views with the Judeo-Christian tradition. When Asa Gray reviewed the *Origin* in 1860, he suggested that Darwin's view of the relation of religion and the natural world was similar to that of the English philosopher William Whewell (1794–1866). Gray's reason for thinking so was that Darwin had placed a quote from Whewell opposite the title page of the first edition of the *Origin:* "But with regard to the material world, we can at least go so far as this—we can perceive that events are brought about not by insulated interpositions of Divine power, exerted in each particular case, but by the establishment of general laws." Gray (1860) suggested in his review what Darwin might have had in mind:

> We judge it probable that our author [Darwin] regards the whole system of nature as one which has received at its first formation the impress of the will of its Author [God], foreseeing the varied yet necessary laws of its action throughout the whole of its existence, ordaining when and how each particular of the stupendous plan should be realized in effect, and—with Him to whom to will is to do—in ordaining doing it. Whether profoundly philosophical or not, a view maintained by eminent philosophical physicists and theologians, such as Babbage on the one hand and Jowett on the other, will hardly be denounced as atheism (182).

Gray's position, following Whewell, was that God created the universe together with the rules that govern the interactions of matter and energy. At the end of creation He went away, leaving events in the natural world to spin out in agreement with His laws. These are the physical laws of mechanics, astronomy, and chemistry that scientists had been discovering since the Scientific Revolution, and evolution through natural selection might also belong on this list. Such a deistic worldview could easily accommodate both science and religion. God is not eliminated, yet nature can still be studied systematically—indeed, scientifically—because God's laws ensure that a given cause acting under defined conditions will always produce the same result. A scientist who seeks to discover the laws of nature will find them, whether he assumes they are ordained by God or are merely the ways that matter and energy naturally behave.

This reconciliation of creationism with science, in which God creates matter, energy, and their governing laws and then retires from the scene—a theological Big Bang, so to speak—was of little comfort to those laypeople who needed to believe in a personal God who was deeply concerned with their daily welfare. And as for scientists, there were other philosophical problems with this deistic worldview that had to be acknowledged. One of these had to do with the principle known as Occam's razor. William of Occam, a highly regarded English monk and philosopher of the fourteenth century, is best remembered for his philosophical position that "entities must not be unnecessarily multiplied." Among scientists, this meant that a minimum number of elements should be used in explaining a given phenomenon. Occam's razor would suggest that in explaining the diversity of life, there is no reason to invoke God to account for the "laws" of natural selection and variability. One could argue that these phenomena were not laws but were rather the inevitable consequences of life itself. This is not to say that God does not exist but only that He is an unnecessary part of the hypothesis.

To this day, people in science and in the Church have continued to wrestle with these problems and to try to adjust the new findings in evolutionary biology to traditional Judeo-Christian beliefs. And the struggle often takes the same forms that it did in the nineteenth century. Among people who believe in God are the strict fundamentalists, who simply deny the data provided by biologists and geologists and persist in believing in a God who created the world in six days and continues to intervene to guide the course of earthly events. A second group of believers might be called separationists; they assign to science the role of explaining the phenomena of nature and to religion the role of providing moral guidance, spiritual expression, and purpose. This position accepts that religion and science deal with different domains and are not in conflict. A third group, the modern deists, believe that God created the world and the laws pertaining to the interactions of matter and energy and since then has let the system run its course without further intervention.

Among those who do not accept any God are two principal groups. The agnostics (T. H. Huxley is credited with introducing this term) maintain that the existence of a God is unknown and unknowable. The atheists, however, deny altogether the existence of a God. Some scientists in this second category are quite vocal in their beliefs, but they should know better: one can no more prove that there is no God than prove that there is one—or more.

Darwin closed the *Origin* with this insightful and moving passage (489–90):

> It is interesting to contemplate an entangled bank, clothed with many plants of many kinds, with birds singing on the bushes, with various insects flitting about, and with worms crawling through the damp earth, and to reflect that these elaborately constructed forms, so different from each other, and dependent on each other in so complex a manner, have all been produced by laws acting around us. These laws, taken in the largest sense, being Growth with Reproduction; Inheritance . . . Variability . . . [and] a Ratio of Increase so high as to lead to a Struggle for Life, and as a consequence to Natural Selection, entailing Divergence of Character and the Extinction of less-improved forms. Thus, from the war of nature, from famine and death, the most exalted object which we are capable of conceiving, namely, the production of the higher animals, directly follows. There is grandeur in this view of life, with its several powers, having been originally breathed into a few forms or into one; and that, whilst this planet has gone cycling on according to the fixed laws of gravity, from so simple a beginning endless forms most beautiful and most wonderful have been, and are being, evolved.

Possibly the most important point here is that these wonders "have all been produced by laws acting around us." Darwin was explaining the diversity of life not as a consequence of supernatural forces but as solely due to the interactions of natural things and processes.

WHAT THE THEORY
OF EVOLUTION EXPLAINS

The probability that a theory is correct becomes greater as it explains more and more data. This was the argument Darwin made. Despite his book's title, he did not provide detailed evidence in the *Origin* for the origin of any species. But he did assemble a vast quantity of data about variations of animals and plants in nature and under domestication, the struggle for existence, natural selection, laws of variation, instincts, hybrids, the paleontological record, geographical distribution, anatomy, classification, and embryology, and showed that numerous puzzles, otherwise inexplicable in natural terms, made sense in the light of evolution.

The Fossil Record

The revolution in geology that Lyell had started several decades before Darwin was the recognition that the strata of sedimentary rocks are arranged in a sequence from oldest to more recent. The absolute ages of the individual strata could only be surmised, but relative age—older or younger—could be established beyond a reasonable doubt. Each major group of strata was found to have a unique population of fossil species. Some of the species might have closely similar counterparts in the strata above or below, and a few cases were known of sequences of slightly different organisms in successive strata.

Both of these phenomena—unique groups of organisms restricted to a single stratum and the apparent slow changes in a single type of organism in successive strata—were confirmed again and again as the nineteenth century progressed. While Genesis had no ready explanation for these data from the rocks, the findings could be easily explained by evolution. In fact, evolution *required* such fossil data. It also required that the time interval from the oldest known rocks to the youngest be immense. The extraordinary differences in the organisms in the oldest strata compared with the youngest strata could not have occurred in just a few millennia. Unknown millions of years were necessary. Scientists did not know then, as we do now, how long it had taken for the strata to form; but their total thickness was known to be many miles, and that amount of deposition would not be possible if creation had taken place only a few thousand years ago.

This history of life through time as revealed by the paleontological record was not absolute evidence for evolution, but evolution provided a satisfactory explanation for it. While the fossil evidence did not actually show the process of change of one species into another—it could not, since fossils are not living and so do not mutate, reproduce, and undergo selection—it did show closely similar fossils occurring in adjacent strata that could be explained best by invoking Darwinian evolution.

Linnaean Hierarchy and the Scala Natura

Many other scientific puzzles concerning anatomy, embryology, classification, and microscopic structure had no satisfactory answers in the absence of the concept of evolution. One of the broadest questions was why species seemed to fall so naturally into Linnaeus's hierarchial groups. Consider, for example, the major group to which human beings belong—the phylum Chordata. This is a heterogeneous group of creatures that includes some marine species—amphioxus and the tunicates—that can look like gelatinous blobs. Most chordates, however, are vertebrates—so-called because they have a series of bones on the dorsal side of the body that form the vertebral column. The major groups of vertebrates are fish, amphibians, reptiles, birds, and mammals. Vertebrates resemble one another in many ways other than just having a vertebral column. For instance, most of their organ systems are similar. The digestive, respiratory, excretory, nervous, skeletal, and reproductive systems are much the same in a trout, a frog, a lizard, a sparrow, and a white rat.

All vertebrates are constructed on the same general plan, that is, they are variations on the vertebrate theme. This astonishing observation could be explained as the consequence of divine creation: the Creator made all the vertebrates as variations on one basic theme. But the same data could also be explained by evolution: one would postulate that a relatively simple vertebrate—a fishlike ancestor—that lived a very long time ago was the progenitor of all vertebrates we see today around us and in the fossil record. Some descendants (modern fishes, whales) became adapted for life in water, others (birds, bats) for a life partially in the air, and still others for all the diverse terrestrial habitats. Nevertheless, all possess a similar basic structural organization of the body because they all descended from a common ancestor.

Figure 3 is a highly schematic representation of evolution as a naturalistic explanation of the origin and diversification of the major groups of chordates. The origin of the chordates is represented at bottom left; the first organisms with the three main characteristics of this phylum, notochord, gill pouches, and a dorsal nerve tube, are thought to have evolved from an invertebrate ancestor. The simplest living chordates, the tunicates and amphioxus, are accepted as the closest living representatives of the most ancient chordates. The seven classes of vertebrates—chordates with a vertebral column—branched off at later times to give rise to many new groups, some of which flourished for long periods and then became extinct while others evolved into the vertebrates still alive. The approximate numbers of recognized living species in each class and examples of each are shown in parentheses. The very approximate dates for the beginnings of each class can be estimated by the numbers, in million years ago (mya), on the vertical axis.

Darwinian evolution also offers a plausible explanation for the *scala natura*. For example, the sequence of vertebrates—fish, amphibians, reptiles, birds, and mammals—observed in the fossil record matches the sequence of increasing complexity of anatomical organization: fish, it turns out, have the least complex bodies, while birds and mammals

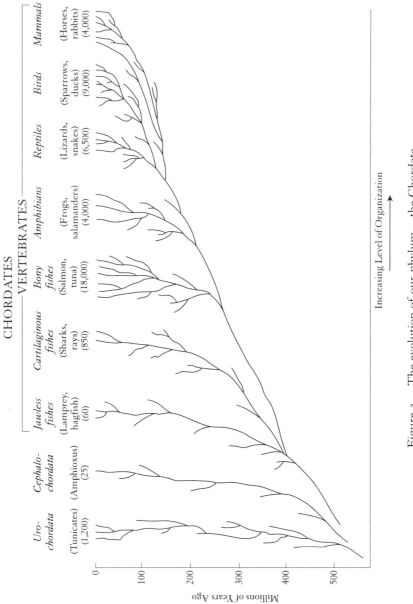

CHORDATES

VERTEBRATES

Uro-chordata (Tunicates) (1,200)

Cephalo-chordata (Amphioxus) (25)

Jawless fishes (Lamprey, hagfish) (60)

Cartilaginous fishes (Sharks, rays) (850)

Bony fishes (Salmon, tuna) (18,000)

Amphibians (Frogs, salamanders) (4,000)

Reptiles (Lizards, snakes) (6,500)

Birds (Sparrows, ducks) (9,000)

Mammals (Horses, rabbits) (4,000)

Increasing Level of Organization

Millions of Years Ago

0

100

200

300

400

500

Figure 3.   The evolution of our phylum—the Chordata.

are the most complex. These parallel results suggest strongly that the underlying cause of the sequence of increasing complexity might be evolution. Both the paleontological data and the anatomical data make sense if we assume that some fishes evolved into amphibians, some amphibians evolved into reptiles, and some reptiles evolved into birds, while other reptiles evolved into mammals. Because the fossil record at the time Darwin was writing was still sparse, naturalists would have thought only that the data were *consistent* with evolution but not absolute proof. Later fossil discoveries would be more and more convincing.

Cellular Makeup of Both Plants and Animals

Another extraordinary observation that evolution explains much better than Genesis does is that the bodies of all animals and plants are composed of the same basic structural unit, the cell. Evolution explains this surprising evidence of relatedness between plants and animals by hypothesizing that the two kingdoms share a common, remote ancestor composed of cells, or perhaps comprising just one cell, like the bacteria and protozoans still living today. In the last part of the nineteenth century it was next to impossible to find fossil evidence for cells, which are microscopic in size and lack hard structures such as bones, teeth, and scales that fossilize readily. But later discoveries have supplied the missing data.

Embryonic Development

The theory of evolution has also proved a powerful tool for explaining some otherwise puzzling aspects of embryonic development. One of the most dramatic and complicated examples concerns the ear bones of all vertebrates except the fishes. The tympanic membrane and the ear bones conduct sound waves from outside the head to the inner portion of the ear, enabling hearing. Fish have neither a tympanic membrane nor ear bones. Amphibians, reptiles, and birds have a tympanic membrane and a single ear bone, the stapes, in each ear. Mammals have a tympanic membrane and three ear bones, the malleus, the incus, and the stapes. If indeed mammals evolved from reptiles, what could be the origin of those two additional bones, the malleus and the incus? (See figure 4.)

The first part of the answer was provided by the German embryologist Karl Reichert in 1837. Reichert found that although the adult mammal has a single bone, the dentary, in the lower jaw, in the embryo there is another bone, the articular. The upper jaw also has an extra bone, the quadrate, which together with the articular in the lower jaw forms the embryonic jaw joint. This condition in the embryo is the *adult* condition in reptiles, birds, and amphibians. Reichert observed that in the course of development the quadrate and articular of the mammalian embryo detach and move to become the malleus and incus of the adult ear. It was suggested, therefore, that in the course of evolution, the quadrate and the articular that formed the articulation of the jaws in other vertebrates had evolved into the incus and malleus of the mammals. Yet how could a switch from one type of joint to another occur? Certainly it could not be instantaneous—evolution is too slow for that. Then could one imagine an intervening form that had two articulations? Could a jaw with two joints really function? Was it possible to obtain evidence to support this hypothesis?

Surprisingly, the answer to these questions proved to yes. Long after embryologists had proposed the transformation of the reptilian jaw bones into the jaws and ear ossicles of mammals, a series of fossils from South Africa and North America revealed that the hypothesis was correct. The fossils belonged to a group of rather early reptiles known as the mammal-like reptiles. And they formed a sequence that showed the conversion of the quadrate and articular into the incus and the malleus, and the transition from the reptilian to the mammalian type of jaw articulation (see figure 4). Once again, the data demanded by theory became available.

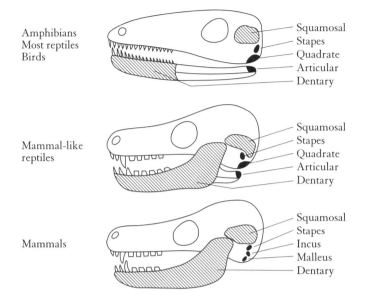

Figure 4. The development of the mammalian jaw and ear bones. (Source: Modified from John A. Moore, *Science as a Way of Knowing,* Harvard University Press, 1993, p. 177.)

Another embryonic puzzle that evolution resolves is the development of the vertebrate kidney. Figure 5 is a schematic representation of the development of the kidneys, which start as tubules linked by a duct that carries urine to the posterior part of the body, where it is expelled. The first kidney to form in the embryos of all vertebrates is the pronephros (derived from the Greek *pro,* meaning "in front," plus *nephros,* meaning "kidney"). The kidney of the adult hagfish, the most primitive living jawless fish known, is thought by some anatomists to be a persisting portion of the embryonic pronephros. The older embryos of all other vertebrates next develop a kidney further back in the body, the mesonephros (*mes* meaning "middle"). The mesonephros remains the kidney of the adult in the lower vertebrates—the cartilaginous, and bony fishes and the amphibians. The embryos of the higher vertebrates develop first a pronephros and then a mesonephros,

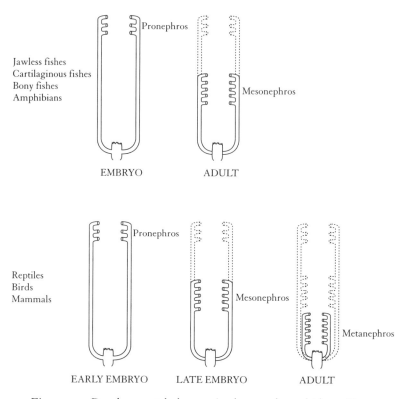

Figure 5.   Developmental changes in the vertebrate kidney. (Source:
Modified from John A. Moore, *Science as a Way of Knowing,* Harvard
University Press, 1993, p. 179.)

but as older embryos form a third kidney, the metanephros (*meta*
meaning "posterior") which remains as the kidney in the adult. Based
on these observations, the hypothesis is that the first vertebrates
probably had only a pronephros and that the higher fishes and amphibians evolved a mesonephros and finally the reptiles, birds, and
mammals evolved a metanephros. The accepted scientific explanation
is grounded in the concept of evolution: the more advanced vertebrates
recapitulate the stages of kidney development in the lower vertebrates.

   The numerous examples of recapitulation in the development of vertebrates, many of which were known in Darwin's time, indicate that some of the early developmental patterns established in the first vertebrates have been retained, though modified, in their descendants. The mammals do not recapitulate all of the structures of the more primitive vertebrates, however. While in the uterus human embryos, for example, do not go through a stage of being covered with fishy scales and swimming around in the amniotic fluid propelled by fins. So why recapitulate a pronephros? The answer to that puzzle had to wait until the twentieth century, when experimental embryologists discovered that if the pronephros is removed from a very early frog embryo, the mesonephros never develops. The pronephros, then, is a necessary though transitory stage for the development of the meso-nephros.

Vestigial Structures

One of the more impressive triumphs of the theory of evolution was to offer a rational explanation for what are known as vestigial struc-tures. A classical example is the human appendix, a small tubular outgrowth of the alimentary canal near the junction of the small and large intestines. It has no known function in digestion and is of interest only when it becomes infected and has to be removed surgically. A study of other mammals has shown, however, that the appendix is an important structure in some species. The rabbit, for example, has a very long and well-developed one in which bacteria live that digest complex carbohydrates, such as cellulose, which make up a large com-ponent of the rabbit's food. Human beings do not have these cellulose-digesting bacteria and thus do not need an elongated appendix to serve as their home, so natural selection has never promoted its development.

There are numerous examples of vestigial structures throughout the animal and plant kingdoms. The ancestors of snakes walked on four legs. Most snakes today have no vestiges of legs, but a few do—tiny bones under the skin where the hind legs should be. Some whales, whose ancestors lived and walked on land, have tiny bones inside the body where the hind legs should be. The whales, however, have retained their front limbs as flippers that aid in locomotion and in maintaining stability. The tail is their main organ of locomotion.

There seem to be few biological puzzles for which Darwinism does not suggest a plausible answer. In some dramatic instances confirmation from the fossil record is possible. And there is nothing in the fossil record that falsifies the theory of evolution. At the close of the nineteenth century, instead of the age-old religious explanations that were still widely held, Darwin offered a useful theory based solely on logical and natural principles: evolution through the mechanism of natural selection acting on variation in a finite environment, resulting in descent with change. Yet evolution is a historical science, which means that very little can be verified by direct observation. The data for ancient organisms, if they persist, lie buried in the geological strata. We cannot replay the tape of the Earth's history and watch dinosaurs evolving into birds. We can do no more than search for indirect data that provide evidence for such a transition.

FOUR

Making the Case
for Evolution

Imagine yourself living in the two decades after the publication of the *Origin*. You have read Darwin's book as well as reviews, heard several lectures on the subject, and possibly even attended the Oxford debate between Wilberforce and Huxley. If you were not a scientist and were unfamiliar with the workings of scientific procedures, you might be somewhat puzzled and feel unsatisfied with the data available to Darwin and other naturalists that were regarded as "proofs" of evolution. You might have thought that evolution meant the conversion of one species into another, as the title of Darwin's book suggested; yet not a single example of this conversion was provided in the *Origin*. Instead, the proofs of evolution consisted of a heterogeneous collection of data: the finding that fossils and living species of armadillos of Argentina are very similar; the restriction of related species in the Galapagos to their own islands; the recapitulation of reptilian jaws in mammalian development; the different kidneys in embryos and adults of vertebrates; vestigial organs such as the human appendix. Would you have found the *Origin* convincing?

The proofs of evolution do not come from the experimental demonstration of one species changing into another. Such direct evidence will always be absent except for unusual situations because the change of one species into another may take many thousands of years—and none of us is able to wait that long for definitive proof. Instead, we have to put together bits of evidence that, with luck, will provide a plausible explanation of what occurred. A detective may study the grooves on the murder bullet, trace the activities of suspects, wonder why the household dog did not bark, compare the DNA in a drop of blood at the murder scene with that of suspects, and seek a motive. Neither the detective not the evolutionist observes the events that require explanation. Both, however, proceed in a similar manner and reach a highly likely conclusion.

The data that prompt biologists and paleontologists to accept evolution as the most accurate statement possible about the diversity of life are indirect, but each piece is consistent with the concept of evolution. Although we cannot "see" evolution on any major scale, much of the data of biology and paleontology cannot be understood without it. Furthermore, the theory of evolution continues to suggest new ways of obtaining further understanding of the origin and diversification of life over the ages. There is no other scientific theory that has proven to be as useful, and for this reason evolution is now accepted as true beyond all reasonable doubt.

Evidence for most of the data that convince scientists is rarely convincing, or even understood, by nonscientists. For example, most people accept what astronomers tell us—that the rotation of the Earth on its axis is the cause of night and day—but can laypeople recite the evidence that this is so? Not likely. To take a more recent example, a layperson who is shown the data on the temperatures in deep space and the change in the spectral lines of light from distant galaxies probably cannot figure out that these data are evidence for the Big Bang that started the universe 12–15 billion years ago. If a nonspecialist accepts the Big Bang as a useful concept, it is because he or she has confidence that astrophysicists are giving the best explanation they can on the basis of available evidence.

There is no strict way scientists—or detectives—go about their quests for answers. There is no one rigid scientific method, but scientists do follow in a general way a series of steps. First, they formulate a question about some natural phenomenon that needs to be explained. The one evolutionists start with is: How can we provide a scientific explanation for the many kinds of organisms that lived in the past and are alive today? The next step is to guess what an answer might be. That guess, or hypothesis, is almost always based on preliminary observations from nature or data from experiments that make the hypothesis plausible. In Darwin's case, he based his hypothesis on his observations of geographic variation among animals and plants in the Galapagos and on the resemblance of fossils in Argentina to living species found there. Both observations could be explained by the hypothesis that evolution works through natural selection acting on the variation present in populations.

The next step is to test the hypothesis to see if it can explain other kinds of data. This is done by deducing the consequences that would occur if the hypothesis were correct. A deduction, then, is a logically necessary derivative from the hypothesis being tested. If the hypothesis is that the reptiles evolved from the amphibians, an important deduction would be that the fossil record should contain individuals that are intermediate between the amphibians and their descendants, the reptiles. Since both are bony creatures, and since bones have a good chance of being fossilized, with enough luck and labor fossils of organisms intermediate between amphibians and reptiles should be discoverable.

If a fossil structurally intermediate between an amphibian and reptile were discovered, the cry "Eureka!" would be premature unless another deduction—that the intermediate organism must have lived *after* the amphibians first appeared—were also found to be true, since descendants cannot be more ancient than their ancestors. To test this deduction calls for reliable methods for determining the age of sedimentary rocks in which fossils are entombed. Just knowing the relative ages of the strata is often satisfactory for this purpose. For example, if all of the intermediates between amphibians and reptiles were in strata younger than those containing the earliest reptiles but older than those with the earliest amphibians, the hypothesis would be substantiated.

A fruitful scientific theory is one that

1.   explains a natural phenomenon with logical and internally consistent arguments

2.   relates that explanation to the existing conceptual scheme

3.   bases that explanation on confirmatory data from observations and experiments

4.   rigorously excludes supernatural phenomena as explanations

5.   reduces the complexity of nature to relative simplicity

6.   is intellectually satisfying and even elegant

7.   suggests experiments and observations that expand the implications of the theory and increase the probability that it is not incorrect

Instead of "increase the probability that it is *not* incorrect," why not "increase the probability that it *is* correct"? The reason is that a single experiment or observation invalidating a scientific statement means that the statement must be modified or abandoned. Suppose the hypothesis to be tested is "Alice is six feet four inches tall and is thus the tallest woman in the world." One could test this hypothesis by comparing Alice with dozens of other women and probably find that she is taller than any. So far so good. But to prove it correct, one would have to measure all of the women in the world. In an actual experiment, it would probably be necessary to check the heights of only several thousand other women to find one taller than six feet four inches. Thus, a single observation would disprove the hypothesis, but an impractical number of observations—checking all women in the world—would be required to prove it correct. A hypothesis in science remains useful if an ever-increasing number of observations seem to indicate that it is correct and no observations prove that it is wrong. Nevertheless, the most accurate comment about any scientific statement is that it not be accepted as true in any final sense but that it be tested repeatedly, never falsified, and so remain for the moment true beyond all reasonable doubt.

This stricture, to never say that a statement in science is true, is philosophically correct, but in the real world such skepticism can border on nonsense. Science does work: rockets do reach the moon and planets; the positions of the major objects in the solar system can be predicted with great accuracy; human diseases can be prevented, cured, or lessened in their severity; and a steady stream of technological wonders emerge from our laboratories and factories, based on knowledge that practically speaking can be accepted as "true." Scientific knowledge may not be true in a philosophical sense, but it is often quite adequate to provide us with an extraordinary level of understanding of ourselves and our world.

THE SEARCH FOR MISSING LINKS

Darwin's theory of evolution implies that although at the microevolutionary level the changes in a lineage are gradual, the summing of these small steps over time results in differences so great that descendants and their ancestors fall into different taxonomic groups: at first, different species, then genera but, with more time and evolution, different families, orders, classes, phyla, and kingdoms.

If this is indeed the way evolution works, surely some fossils intermediate in structure between two major groups, such as fishes and amphibians or reptiles and birds, could be found. In Darwin's day these intermediates were called missing links because for so long they were indeed missing. Darwin could not provide a single example of this kind of intermediate, such as between two classes of vertebrates. Needless to say, the antievolutionists argued that the absence of this critical information made the hypothesis of evolution improbable. Strict creationists continue to make this argument today, in spite of the many fossils intermediate between major taxonomic groups that have been discovered.

The first, and in many ways still the most spectacular, discovery of a fossil intermediate between major groups was *Archaeopteryx* ("primitive bird"), found in 1861, just two years after Darwin published. It was discovered in the famous quarry near Solenhofen, Germany, where a fine-grained limestone was mined because it was especially good for making lithographic plates. The limestone is a stratum of the Jurassic period that is about 140 million years old. The first birdlike fossil discovered was only the impression of a single feather. It was in strata far older than any known at the time to contain birds. How wonderful and frustrating—birds must have been present, since no other class of vertebrates has feathers, but what were they like? A few months later an entire specimen was discovered and was eventually bought by the British Museum. A total of seven specimens of *Archaeopteryx* are now known.

Careful study showed that *Archaeopteryx,* which was about the size of a pigeon, had a blend of reptile and bird characteristics. Its feathers are characteristic of birds, but its jaws and teeth are reptilian; modern birds do not have teeth. The upper portion of the skull is birdlike, and the front appendages are modified as wings. But unlike modern birds, three of the fingers are not fused, and they end in claws. A birdlike wishbone is present, but the sternum has no keel, a structure characteristic of most modern birds. The absence of a keel suggests that *Archaeopteryx* was a glider, not a true flier, because the keel in modern birds serves as the attachment for the strong muscles that flap the wings. And finally, *Archaeopteryx* had a very reptilian long tail.

Thus, *Archaeopteryx* is a mosaic of reptilian and avian characteristics and as intermediate between two major groups of organisms as one could desire. However, a choice regarding its category of classification had to be made for practical reasons. Because its body was covered with feathers, taxonomists decided that *Archaeopteryx* should be classified as a bird. Had only the skeleton been preserved, *Archaeopteryx* would have been identified as a reptile. Indeed, exactly that error had been made before *Archaeopteryx* was discovered. Some fossils consisting only of bones were so similar to small dinosaurlike reptiles that they were classified as reptiles. Later, it was found that those bones were identical to the ones in the complete skeleton of *Archaeopteryx* and were, therefore, fossils of *Archaeopteryx*. The bodies of these individuals must have undergone considerable decay before being covered by silt, so that only bones—no feathers—remained to be fossilized. This is a dramatic illustration of the intermediate nature of *Archaeopteryx:* without feathers, they could be considered reptiles; with feathers, they are considered birds.

Evolution is a seamless strand devoid of major changes in any brief span of time. What appear as major changes are in fact artifacts in our knowledge usually due to the absence of an adequate series of fossils. When relatively few fossils of the reptiles that were the putative ancestors of birds were known, the structural gap between reptiles and *Archaeopteryx* seemed pronounced—but only because of the paucity of data. Those data are now becoming available. Numerous fossils are being discovered that are closing the gap between advanced reptiles and the earliest birds. It is clear that the many characteristics that distinguish reptiles from birds did not change all at once. One would predict, therefore, that there must have been a long period during which the evolving population could not be called either reptile or bird. At no point could one draw a line and say that every generation up to this point is reptilian and that all subsequent generations are birds. For that to be possible one would have to choose one parental generation to be labeled "reptiles" and call its children "birds." Paleontologists today are digging up the evidence for the transition period and talking about dinosaurs with feathers.

About a decade after the discovery of *Archaeopteryx,* two very different species of toothed birds were discovered in deposits from the Cretaceous period in the western United States. One was named *Hesperornis* ("place of sunset," or "western," plus "bird") and the other *Ichthyornis* ("fish bird"). *Hesperornis* was very large, standing about three feet tall. The wings, however, were small, and the form of its body suggested that it was a diving bird, similar to the loons of today. *Ichthyornis* was gull-like in size and possibly in habits. In contrast to *Archaeopteryx,* it had a well-developed keel on the breastbone that suggested it was a strong flyer. Both genera had teeth, which strengthened the evidence that birds had descended from reptiles.

These Cretaceous birds were collected by Othniel Charles Marsh of Yale University and were described by him in a classic of paleontology, *Odontornithes: Monograph on the Extinct Toothed Birds of North America* (1880). As is usually the case, there are fascinating stories behind these early collecting expeditions in the American West. Settlements were few west of the Mississippi River, and travels beyond could be difficult and dangerous. Some of the flavor of what it meant to be a paleontologist in those times is provided by this description in Marsh's introduction.

> The first Bird fossil discovered in this region was the lower end of the tibia of *Hesperornis,* found by the writer in December, 1870, near the Smoky Hill River in Western Kansas. Specimens belonging to another genus of the *Odontornithes* were discovered on the same expedition. The extreme cold, and danger from hostile Indians, rendered a careful exploration at that time impossible.
>
> In June of the following year, the writer again visited the region, with a larger party, and a stronger escort of United States troops, and was rewarded by the discovery of the skeleton which forms the type of *Hesperornis regalis.*
>
> Although the fossils obtained during two months of exploration were important, the results of this trip did not equal our expectations, owing in part to the extreme heat (110° to 120° Fahrenheit, in the shade) which, causing sunstroke and fever, weakened and discouraged guides and explorers alike. (2)

The primitive birds found in Europe and America provided the first good evidence for the intermediates between major vertebrate groups that the theory of evolution requires. Today we have many more examples. Fossils are now known that are intermediate in their characteristics between fishes and primitive amphibians and between amphibians and primitive reptiles. One of the better-documented cases is that of the fossils of the mammal-like reptiles of the Triassic period.

Can intermediate types be considered ancestors? For instance, were *Archaeopteryx* and the other extinct fossil birds the ancestors of modern birds? Many people in the late nineteenth century jumped to that conclusion. However, today one would be hard pressed to find a paleontologist willing to say so. Rather, researchers today are careful to recognize the limits, of their methods and only say that there were probably many different species of birds living when *Archaeopteryx* was alive, and one or more of these must have been the ancestors of later types. It is also quite possible that there is no present-day bird for which *Archaeopteryx* was a direct ancestor.

The fossil record for the vertebrates is now adequate to reveal the broad outlines of the evolution of the major groups: fish, amphibians, reptiles, birds, and mammals (see chapter 3, figure 3). A great deal is also known about evolution within every group except the birds. The data are especially good for mammals and reptiles. Mammals have the advantage of a fairly recent origin and so a shorter period for their fossil record to be destroyed, and their skeletons and teeth provide excellent material for determining lineages. Teeth are especially useful because they vary greatly among the different groups of mammals. Some paleontologists have said, only partly in jest, that it is possible to determine the evolutionary history of mammals from their fossil teeth alone. The teeth of fish, amphibians, and reptiles are less useful because they are fairly uniform, and modern birds are toothless.

The mammals also provide some good evidence regarding evolution within families, the horses being a classic example. In 1874 Marsh published an early version of the story of the evolution of the horse, based on fossils collected on his western expeditions:

> The animals of this group which lived in this country during . . .
> the Tertiary period were especially numerous in the Rocky Moun-

tain regions, and their remains are well preserved in the old lake basins which then covered so much of that country. The most ancient of these lakes—which extended over a considerable part of the present territories of Wyoming and Utah—remained so long in Eocene times that the mud and sand, slowly deposited in it, accumulated to more than a mile in vertical thickness. In these deposits, vast numbers of tropical animals were entombed, and here the oldest equine remains occur. (288)

But Marsh found more than the earliest fossil horses. He found four fossil species, each restricted to a different layer of Tertiary rocks but together spanning almost the entire Tertiary period. Today geologists divide the Tertiary period into five epochs, from Paleocene through Pliocene (see figure 10, p. 113). There follows the very brief Quarternary period, including the Pleistocene and Recent epochs, the latter a mere ten thousand years. It was in the Recent epoch that human beings became a dominant force on Earth.

The first fossil horse Marsh discovered was *Orohippus,* from the Middle Eocene. Several years later, however, he found *Hyracotherium* from the Early Eocene, which remains the most ancient known genus of the horse family (see figures 6a and 6b). Next in order of descent was *Miohippus* from the Miocene, and the youngest fossil was *Hipparion,* from the Pliocene. The modern horses belong to the genus *Equus.*

Thus Marsh could consider five genera that gave a view of the history of horses over the last 50 million years. To be sure, an average of one ancient horse for every 10 million years cannot be considered a very complete record, but at least it was a beginning. The differences in the anatomy of these horses were not random but seemed to exhibit trends that allowed one to understand how the *Hyracotherium* structures gradually changed into those of other genera to finally produce the modern horses of the genus *Equus.* The two Eocene genera, *Hyracotherium* and *Orohippus,* were very small animals, about the size of a fox or small dog. The size of horses increased during the Tertiary

Figure 6a.   An illustrated reconstruction of the earliest known horse, *Hyracotherium,* of the Eocene epoch. (Neg. no. 35767; photo by AMNH Photo Studio. Courtesy of Department of Library Services, American Museum of Natural History.)

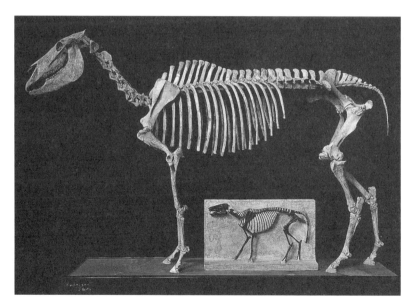

Figure 6b.   The skeleton of the diminutive *Hyracotherium* is nestled beneath that of *Equus,* the genus of present-day horses. (Neg. no. 35282. Courtesy of Department of Library Services, American Museum of Natural History.)

period, *Miohippus* (from the Miocene) and *Hipparion* (from the Plio-cene), each being progressively larger. In fact, *Hipparion* was as large as some of the smaller breeds of contemporary horses.

Of exceptional interest is the evolution of the structure of the horses' front legs (see figure 7). The more primitive Paleocene mammals from which horses presumably evolved had five digits on each appendage, as is the case with our hand and foot. The Eocene horses, however, had four toes on the front feet and three toes on the hind feet, each with a tiny hoof. Counting the innermost toe of a five-digit appendage (like our thumb or big toe) as number I, the front legs had lost number I and only numbers II, III, IV, and V remained. The hind foot had lost toes number I and V. The discovery of little horses with three or four toes, each with a little hoof, proved fascinating not only to pale-ontologists but to the general public as well. Both Miocene *Miohippus* and *Hipparion* in the Pliocene had three toes on the front feet, but the side toes of *Hipparion* were smaller than the other toes. Modern horses have a single toe, number III, on each foot, and the vestigial remains of toes II and IV are present as the splint bones, as shown in Marsh's illustration. All horses of today walk around on what were once the third toes of their ancestors. Even before these fossil horses were dis-covered, comparative anatomists had concluded that the modern horse had a single functional toe and the splint bones were vestigial toes. The discovery of fossil horses validated their hypothesis.

There were many other changes in the skeleton, including a great increase in size from the fox-size dimensions of *Orohippus* to *Equus* today. The lengthening of the limb bones permitted increased running speed. The neck and the front part of the skull—the area in front of the eyes—became elongated.

Horses remain one of the most interesting examples of the evolution of a vertebrate family. In spite of the fact that today North America has no native horses, it was the center of horse evolution. A worldwide total of about 34 genera of fossil horses have been discovered, and all became extinct except one, *Equus,* which includes horses, zebras, asses

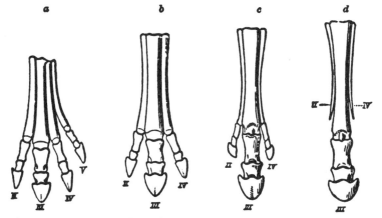

a, Orohippus (Eocene); *b*, Miohippus (Miocene); *c*, Hipparion (Pliocene); *d*, Equus (Quaternary).

Figure 7.   Marsh's original drawing illustrating the evolution of
the horse's feet (showing a front foot) begins with the four-toed genus
of the Eocene, *Orohippus,* and continues with the Miocene *(Miohippus)*
and Pliocene *(Hipparion)* genera. The last shows the modern
horse, *Equus.*

(donkeys), and onagers. Although having 34 genera to represent about
50 million years of evolution might not seem like a paleontological
cornucopia, the relative abundance of individual fossils has provided
considerable information about the habits and distribution of the
members of the horse family.

The structure of the teeth suggests that the horses that lived before
the Miocene were browsers (that is, they ate the leaves of trees) rather
than grazers (which eat grasses and other low vegetation). This change
occurred over a period of a few million years across many different
genera of horses and can be attributed to a worldwide climate change.
The gradual cooling and drying of the climate during the Miocene led
to a decrease of forests and an increase in grasslands, which prompted
the horses to switch from browsers to grazers worldwide. That introduced a new problem. The silicon in grasses is a very abrasive sub-

stance that causes wear on the teeth. Correlated with the new environmental challenge, natural selection promoted a change in the teeth. The adult teeth of other mammals, such as human beings, do not grow after they appear, but those of horses and many other grazing animals continue to grow throughout life; they thus do not become worn down to the gums but remain full-sized.

Nearly all the Miocene genera of horses became extinct, and not many were left during the Pliocene; but *Equus* appeared in North America, and several genera migrated to South America. *Equus* also migrated to the Old World, and various species evolved in Europe, Asia, and Africa. All New World horses were extinct by about 11,000 years ago, when the Pleistocene Ice Age ended. The cause is unknown, but it was part of a mass extinction that saw the death of many other large mammals, an estimated 186 species in all. The New World mammoths, mastodons, camels, giant sloths, giant bears, and sabertooth cats vanished. Horses survived only in the Old World. Today's "wild" horses of North and South America are descendants of domesticated European horses introduced by the Spanish in the sixteenth century.

The horse story became known only a decade after the discovery of *Archaeopteryx*, and it suggested the same general process—evolution. It is important to emphasize, once again, the nature of the evidence. The sequence in geological time of *Hyracotherium, Orohippus, Miohippus, Hipparion,* and *Equus* shows a series of intermediate stages that link the Eocene horses, *Hyracotherium*, with the living species of *Equus*. But a careful paleontologist would not say, for instance, that *Miohippus* was the ancestor of *Hipparion;* only that it could have been. The possibility that another unknown genus very similar to *Hipparion* was the true ancestor cannot be dismissed. The concept of evolution is supported by the discovery of intermediate fossils, not "true" ancestors. Thus, the hypothesis would be that if *Equus* evolved from some Eocene horse through a long line of intermediates, the paleontological record should contain fossils intermediate between *Hyracotherium* and *Equus*. Diligent and lucky digging might reveal these intermediates.

Marsh was both diligent and lucky. The evidence for a direct ancestor to a descendant would require an enormous number of fossil specimens from closely similar times.

The story of horse evolution was brought to the American public by the outstanding English naturalist and vigorous defender of Darwinism, Thomas Henry Huxley. On September 20, 1876, two years after Marsh had published his results, Huxley gave a series of public lectures in New York City. He had visited Marsh at Yale University and learned firsthand the story of fossil horses in the American West. His third lecture, "The Demonstrative Evidence of Evolution," was about the evolution of horses' toes.

> That is what I mean by demonstrative evidence of evolution. An inductive hypothesis is said to be demonstrative when the facts are shown to be in entire accordance with it. If that is not scientific proof, there are no merely inductive conclusions which can be said to be proved. And the doctrine of evolution, at the present time, rests upon exactly as secure a foundation as the Copernican theory of the motions of the heavenly bodies did at the time of its promulgation. Its logical basis is precisely of the same character— the coincidence of the observed facts with theoretical requirements.
>
> The only way of escape, if it be a way of escape, from the conclusions which I have just indicated, is the supposition that all these different equine forms have been created separately at separate epochs of time; and, I repeat, that of such an hypothesis as this there neither is, nor can be, any scientific evidence; and, assuredly, so far as I know, there is none which is supported, or pretends to be supported, by evidence or authority of any other kind. I can but think that the time will come when such suggestions as these, such obvious attempts to escape the force of demonstration, will be put upon the same footing as the suppositions made by some writers, who are, I believe, not completely extinct at present, that the fossils are mere simulacra, are no indications of the former existence of the animals to which they seem to belong; but

that they are either sports of Nature [what we would now call mutations], or special creations, intended—as I heard suggested the other day—to test our faith.

In fact, the whole evidence is in favor of evolution, and there is none against it. And I say this, although perfectly well aware of the seeming difficulties which have been built up upon what appears to the uninformed to be a solid foundation. (Huxley 1877, 90–91)

Huxley's analysis was accurate when he spoke to that New York audience in 1876 and remains so today. Surely that would please him. But he would be discouraged to find that the same sorts of criticisms of evolution current in his time are still accepted today by the uninformed and by those who reject the statements of science if they conflict with religious beliefs.

RADIOACTIVE DATING

What if Marsh had set up camp in a promising area having cliffs with thick layers of stratified rocks; each of four of his assistants had returned with a different fossil horse that would in time be named *Hyracotherium, Orohippus, Miohippus,* and *Hipparion;* and close examination of the fossils had revealed that the four genera seemed to form a series of increasing size and complexity. Could one reasonably conclude that these four ancient horses were an evolutionary series? Not unless it was found that the four fossils came from four different geological strata that matched in order, from oldest to youngest, the order of the fossils from smallest and simplest to largest and most complex.

Older and *younger* are adequate for assessing relative age, but clearly an accurate method of determining exact geological ages was needed. Various methods were suggested for determining absolute ages but no acceptable method was found. That sad state of affairs is illustrated by the range of dates proposed for the beginning of the Cambrian period—when the first chordates came on the scene—ranging from 70 million years to 6,000 million years ago.

As is so often the case in scientific research, a solution in one field comes from discoveries in another field. In the case of how to tell geological time the solution came from physics. It was radioactivity that provided a method for geologists to obtain confirmable evidence for the true age of a fossil.

Before the 1890s atoms were defined as the ultimate particles of matter—the smallest physical objects in existence and indivisible (the word *atom* is from a Greek word meaning "indivisible"). Atoms were known to differ from element to element, but all atoms of a given element were thought to be identical. This was a reasonable deduction, since chemical reactions between the same elements always produced the same product or products. One oxygen and two hydrogen atoms always combine to form water, and for all intents and purposes water is water.

During the 1890s three scientists in France—A. H. Becquerel, Polish-born Marie Curie, and her husband, Pierre Curie—embarked on a series of experiments that would revolutionize physics, chemistry, biology, and to some extent geology. Physicists were beginning to suspect that atoms were composed of smaller particles, that they were neither indivisible nor constant, and that elements could change into other elements under certain conditions. Such changeable elements are said to be radioactive: in the process of changing from one element to another, they emit subatomic particles. For example, radioactive uranium goes through a series of about a dozen steps in which it becomes thorium, protoactinium, radon, polonium, bismuth, and other elements until it finally ends up as lead. A striking feature of this radioactive decay—and the one that makes it useful to geologists—is that the rate of this radioactive decay is constant and cannot be influenced by any known physical condition, such as temperature or pressure. This is why an atomic clock, which measures the rate of disintegration of radioactive elements, is amazingly accurate.

The time required for half of the atoms in a radioactive sample to reach a nonradioactive state is called the element's half-life. In the case of uranium, the half-life is the interval required for half of a sample of uranium to become lead; scientists determined that this takes roughly 4.5 billion (4.5×10^9) years. Thus the ratio of uranium to lead in a sample can be used to determine when the reactions began. For example, if a sample of rock had equal amounts of uranium and lead, that rock would have been formed 4.5 billion years ago. Great care must be taken in selecting the sample of rock to be analyzed. A most important requirement is that no loss of the breakdown products of uranium should have occurred; they must be trapped in the rock and measured. If any portion is lost, the estimate will be incorrect.

How can one be sure the lead in the sample of rock being tested came from the breakdown of uranium and was not just a normal part of the rock? The answer involves the notion of isotopes. By the time the uranium method for dating rocks was being developed, it had been discovered that all of the atoms of an element need not be identical; they could differ in the number of subatomic particles of which they are composed. These different versions were called isotopes. There are several isotopes of uranium; the one used for age determination is ^{238}U, the 238 being its atomic mass. When this uranium isotope disintegrates, it ends as a rare isotope of lead known as ^{206}Pb. The common isotope of lead, which is present in many rocks, is ^{207}Pb. The uranium method for determining the age of a rock cannot work unless one can distinguish ^{206}Pb from ^{207}Pb. Methods are available to do this with a high degree of accuracy. Therefore age can be determined by the ratio of ^{238}U to ^{206}Pb.

Today many different techniques to date materials are based on radioactivity. The radioactive isotope most often used in dating recently living objects is carbon 14 (^{14}C). It is especially useful for dates in human prehistory because it has a very short half-life—5,700 years. In carefully prepared samples it can be used to date carbon-containing material, such as wood, that is as old as 70,000 years.

Different isotopes can be used to date materials from longer spans of geological time—for example, the conversion of potassium to argon (half-life of 1.2×10^9) and rubidium to strontium (half-life of 50×10^9). Whenever two methods are used to date the same sample and they give the same answer, we can be more confident that the answer is correct.

With radiometric methods available for accurately dating sedimentary rocks, geologists could estimate the ages of strata with increasing exactitude. The great age of the Earth and the vastness of the time that life has been evolving came as a great shock to some. The age at which our solar system, including the Earth, began to form from cosmic dust is now estimated at about 4.5 billion years ago (see figures 8–10)—a far cry from 4004 B.C., Bishop James Ussher's estimate of the year God created the heavens and the earth. This new value is based mainly on radioactive dating of meteorites, which formed at the same time as the sun and planets. Geologists do not base estimates of the age of the Earth on radioactive dating of Earth's own rocks, because at its beginning the Earth was a hot, molten mass, and the oldest rocks could have been formed only after the Earth had cooled. The oldest rocks known on Earth have been dated to about 3.8 billion years old. One can learn from these very old rocks something about the composition of the Earth's crust and its atmosphere at that time, and with luck, possibly find evidence of any life that might have existed then.

LIFE OVER TIME

Throughout the nineteenth and early twentieth centuries, there was no good fossil evidence for life before the Cambrian period, which began about 540 million years ago (see figure 9). It seemed that in the Cambrian period an impressive variety of animals had suddenly appeared— complex representatives of most of the major phyla that are still alive today. How could one account for the sudden appearance—on a geological time scale—of such relatively complex animals as shellfish,

Figure 8.   Ancient times: from the Big Bang to the present.

GEOLOGICAL EONS	BILLIONS OF YEARS AGO	
Phanerozoic	0	Present
		The Phanerozoic, the final eon of Earth's history, witnessed the appearance of a rich variety of complex animals and plants about 540 million years ago. Refer to figure 9.
	1	The oldest known animal fossils date to about 650 million years ago.
Proterozoic		The earliest known multicellular organisms date to about 1.7 billion years ago.
	2	Cells with nuclei are first known from about 2 billion years ago.
Archean	3	
	4	The Archean begins about 3.8 billion years ago, when the Earth's crust became solid, ensuring that a paleontological record could begin to form. The earliest evidence of life, similar to very simple bacteria, dates to about 3.5 billion years ago.
Hadean		Our solar system, consisting of the sun, planets, asteroids, and comets, was forming about 4.5 billion years ago. The Earth was a molten mass that continued to increase in size as meteors crashed into it. By the end of the Hadean eon, the Earth had cooled enough for liquid water to be present. Life was becoming possible.
	5	
	6	
	7	
	8	
	9	
	10	Our celestial home, the Milky Way Galaxy, was born about 10 billion years ago. It now contains an estimated 10 billion stars, of which about 6,000 are visible to the unaided eye.
	11	
	12	The slow cooling of the universe permitted more complex atoms to form. Gravitational forces started to clump these atoms as cosmic dust, galaxies, and stars. Stars have a cycle of birth, maturity, and finally death.
	13	
	14	
	15	The Big Bang of about 15 billion years ago started the universe. It was a huge explosion that sent matter, consisting of subatomic particles, including protons, electrons, neutrons, and others, flying out in all directions. The temperature was about 100 billion degrees Kelvin. When the universe began to cool, the subatomic particles combined to form the smallest atoms, such as hydrogen and helium.

Figure 9.   Ancient times: the past 600 million years.

ERA	PERIOD	MILLIONS OF YEARS AGO	
Cenozoic	Quarternary	0	In the most recent 2 million years of Earth's history ice ages dominated the northern continents. Human beings spread worldwide from Africa. Civilization began. Refer to figure 10.
Cenozoic	Tertiary	65	The vacuum left by the mass extinctions at the end of the Cretaceous permitted the rapid evolution of mammals, birds, and terrestrial life in general. Mammals, not reptiles, dominated the landmasses. Grasses and grazing animals became abundant. The primates evolved from the very primitive kinds to the level of monkeys and advanced apes. Refer to figure 10.
Mesozoic	Cretaceous	100 / 144	Reptiles continued to be the dominant vertebrates. Flowering plants, birds, and mammals were present but not conspicuous. At the end of the Cretaceous, there was a widespread extinction of life probably caused by a huge meteor striking the Earth.
Mesozoic	Jurassic	200 / 206	The reptiles continued their extensive evolutionary radiation on land and sea and in the air. Birds and mammals appeared.
Mesozoic	Triassic	248	The mass extinction of life that ended the Permian provided an opportunity for the reptiles to begin an incredible evolutionary radiation. Pangea began to break into the continents, which began to move to the positions they occupy today.
Paleozoic	Permian	290	The comparatively primitive fauna and flora of the Paleozoic underwent the Earth's greatest mass extinction due to intense cold at the end of the Permian. The landmasses joined to form a single continent, Pangea. Mammal-like reptiles were present.
Paleozoic	Carboniferous	300 / 354	Extensive growths of primitive plants in the warm, humid conditions produced extensive deposits of coal. Amphibians were the dominant land vertebrates, and by the end of the period some had evolved into reptiles.
Paleozoic	Devonian	400 / 417	This was the age of fishes. There were no other vertebrates until the end of the period, when some fishes evolved into amphibians. The land had become colonized by primitive plants and many arthropods. A massive extinction of life occurred at the end of the period.
Paleozoic	Silurian	443	A major event was the beginning of life on land, accomplished by arthropods, such as scorpions, mites, and spiders, and primitive green plants.
Paleozoic	Ordovician	490	Many new forms of life evolved and replaced the many kinds that had died at the end of the Cambrian. All life was still in the seas. Primitive fishes, the first vertebrates, appeared.
Paleozoic	Cambrian	500 / 543	This is the first period with a variety and abundance of life, including all animal phyla. This "sudden appearance" is correlated with the presence of tough exoskeletons that fossilize well. Primitive chordates, but no vertebrates, occurred. Landmasses began to move together to form Pangea in the Permian. A massive extinction of life occurred at the end.
Proterozoic	Vendian	600 / 670	The Precambrian world, essentially unknown until recently, was one of very simple animals, with uncertain relations to the Cambrian fauna, and a variety of algae. The Ediacarian organisms, first discovered in Australia, are now known to have been widespread. The environment was terrible for life: long periods of intense cold probably exterminated most creatures.

starfish, corals, and crustaceans? One hint of the answer is that all of the Cambrian animals known at the time had shells or other hard structures. Soft body parts such as internal organs very rarely fossilize, so animals in the Precambrian, which had not yet evolved complex shells, would not have been so readily preserved. The animals of the Cambrian only *appeared* to have suddenly shown up; there is now abundant evidence that these complex animals were the results of a long period of evolution in the Precambrian.

In recent decades special techniques have been developed for searching very ancient sedimentary rocks for evidence of soft-bodied animals and even simple bacteria. Before these techniques were available, it was far more exciting to search for the noble dinosaurs than for microscopic creatures with the simplest anatomy. But by the middle of the twentieth century the general patterns of animal evolution were fairly well known, and the question of what happened in those vast eons before the Cambrian became a prime topic for research. As a result of these investigations, the date for the first evidence of life has been pushed back to about 3.5 billion years ago, shortly after the first rocks formed on Earth (see figure 8, p. 110). Life at the beginning consisted of primitive bacteria, some of which cannot be distinguished from the prokaryotes living today.

The prokaryotes were the only forms of life in the interval from about 3.5 to about 2 billion years ago when eukaryotes came on the scene. Both the prokaryotes and the eukaryotes have the same hereditary material, DNA, but in the prokaryotes it lies free in the cell. In eukaryotes most of the DNA is enclosed in a vesicle called the nucleus. (DNA is also found in tiny organelles within cells, called the mitochondria, and in the chlorophyll of green plants.) From 2 billion to about 650 million years ago there is no generally accepted fossil evidence for animals that are composed of many eukaryotic cells—the multicelled animals, or metazoans. Recent discoveries have found metazoans beginning about 650 million years ago. By the onset of the Cambrian, 540 million years ago, metazoans appeared in abundance.

Figure 10. Ancient times: the past 65 million years.

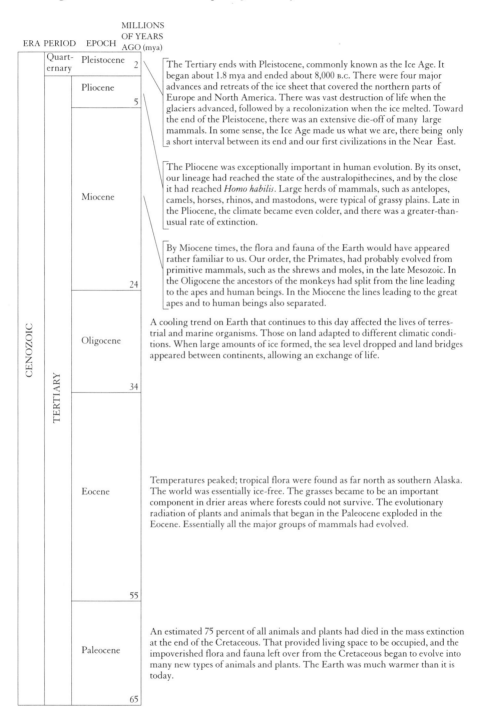

ERA	PERIOD	EPOCH	MILLIONS OF YEARS AGO (mya)	
CENOZOIC	Quart-ernary	Pleistocene	2	The Tertiary ends with Pleistocene, commonly known as the Ice Age. It began about 1.8 mya and ended about 8,000 B.C. There were four major advances and retreats of the ice sheet that covered the northern parts of Europe and North America. There was vast destruction of life when the glaciers advanced, followed by a recolonization when the ice melted. Toward the end of the Pleistocene, there was an extensive die-off of many large mammals. In some sense, the Ice Age made us what we are, there being only a short interval between its end and our first civilizations in the Near East.
		Pliocene	5	The Pliocene was exceptionally important in human evolution. By its onset, our lineage had reached the state of the australopithecines, and by the close it had reached *Homo habilis*. Large herds of mammals, such as antelopes, camels, horses, rhinos, and mastodons, were typical of grassy plains. Late in the Pliocene, the climate became even colder, and there was a greater-than-usual rate of extinction.
		Miocene	24	By Miocene times, the flora and fauna of the Earth would have appeared rather familiar to us. Our order, the Primates, had probably evolved from primitive mammals, such as the shrews and moles, in the late Mesozoic. In the Oligocene the ancestors of the monkeys had split from the line leading to the apes and human beings. In the Miocene the lines leading to the great apes and to human beings also separated.
	TERTIARY	Oligocene	34	A cooling trend on Earth that continues to this day affected the lives of terrestrial and marine organisms. Those on land adapted to different climatic conditions. When large amounts of ice formed, the sea level dropped and land bridges appeared between continents, allowing an exchange of life.
		Eocene	55	Temperatures peaked; tropical flora were found as far north as southern Alaska. The world was essentially ice-free. The grasses became to be an important component in drier areas where forests could not survive. The evolutionary radiation of plants and animals that began in the Paleocene exploded in the Eocene. Essentially all the major groups of mammals had evolved.
		Paleocene	65	An estimated 75 percent of all animals and plants had died in the mass extinction at the end of the Cretaceous. That provided living space to be occupied, and the impoverished flora and fauna left over from the Cretaceous began to evolve into many new types of animals and plants. The Earth was much warmer than it is today.

The Precambrian strata are now being intensively searched for evidence of soft-bodied multicelled animals and with considerable success. The oldest group of soft-bodied metazoans to be discovered to date is in Australia (see chapter 5), but others are turning up in other parts of the world.

Looking at our own phylum, the Chordata, the fossil record shows that the oldest chordate so far discovered comes from Cambrian rocks. It was a soft-bodied animal of very simple structure, belonging to a group of marine chordates that includes amphioxus and the tunicates, both of which are alive today. It is assumed that a chordate of this low level of complexity was the ancestor of the vertebrates.

In the Ordovician period, with the advent of the vertebrates, whose skeletons fossilize readily, the evidence for evolution vastly improves. The first vertebrates, as we have seen, were primitive fishlike animals. Fishes became especially abundant in the Devonian period, which is commonly called the Age of Fishes. The amphibians, and shortly afterward the reptiles, appeared in the Carboniferous period. The entire Mesozoic era is known as the Age of Reptiles, when the reptiles dominated the land, air, and to some degree the seas. The first known birds and mammals date to the Jurassic era.

The impressive data on the evolution of the vertebrates, as well as similar evidence from other groups of animals and some plants, have convinced both scientists and nonscientists familiar with those data that the concept of evolution is a satisfactory explanation of the present diversity of living organisms as well as the variations that occurred over time and are entombed in the fossil record. That being the case, it may come as a surprise to learn that the study of evolution is more intense today than at any other time in the past. The reason is a familiar one in science: when new techniques and new theories become available, new data and new perspectives emerge rapidly. The new techniques today are those developed by molecular biologists. Molecular data are now available that throw light on the classification of organisms, the degrees of relatedness of different species, the times at which evolutionary lineages split, and rates of evolution.

In addition, ancient rocks are being studied for any organic molecules they may contain, suggesting that life was present. Other studies provide information about ancient climates. The movements of continents have been studied extensively, and this information throws much light on migrations and the geographic distribution of animals and plants. Methods for studying fossils of bacteria are providing information about the simplest kinds of life that dominated the world for billions of years—and in some ways still do. The Precambrian— once a dark period in our knowledge of the past—is finally opening up to scientific scrutiny. Not only can new questions about evolution be asked and answered, but many solutions to age-old puzzles about the secrets of life are being unlocked from the Earth's rocks for the first time.

FIVE

Twentieth-Century Evidence

Darwin's magisterial assembling of the indirect evidence for evolution was convincing to many scientists almost from the publication of the *Origin,* and by the end of a decade to nearly all. However, his hypothesis for the *mechanism* of evolutionary change, namely, inherited variations acted upon by natural selection in a finite environment, was not so convincing. A major reason for this was the near total lack of information in Darwin's time about the origin of variation and the workings of inheritance.

THE RULES OF INHERITANCE

In the last half of the nineteenth century it was not clear whether or not inheritance is a constant and repeatable phenomenon. Some observations suggested that there might be rules of some sort for inheritance, but equally persuasive evidence suggested a fickleness at best. It was obvious, of course, that the offspring of human beings were human beings and that this principle applied to all known life—the species of animals and plants "breed true." It was equally obvious that children of the same human parents were far from identical except in the rare cases of identical twins. Sometimes parental characteristics seemed to be inherited, but in other cases, not. One of the most confusing of all was that a parental feature would not be expressed in the children but would reappear in the grandchildren. And then there was that most astonishing difference of all: offspring of the same parents can be either males or females, which differ greatly in structure, physiology, behavior, and reproduction. In most species that reproduce sexually, about half the offspring are females and the others males. It is obviously convenient that approximately equal numbers of males and females are born, but convenience is not a mechanism—so what could possibly account for this phenomenon?

Darwin's theory required a pool of inherited variations among individuals of the same species from which natural selection could choose those that were more fit for survival and reproduction. When Darwin published the *Origin,* he believed that inheritance must have a definite biological basis and that, whatever it was, it had to provide offspring like the parents but with minor variations. There were no data to explain how this was possible during the last half of the nineteenth century. In subsequent editions of the *Origin* Darwin was driven more and more to the belief that the variations he saw in nature were due in some degree to the direct action of the environment. This had been the suggestion of Lamarck back at the beginning of the century, and by the century's end a belief in Lamarckism had become common among Darwin's contemporaries.

The modern understanding of inherited variations began in 1900, more than a decade after Darwin's death. In that year, experiments on inheritance in garden peas conducted in the 1860s by an Austrian monk and amateur naturalist, Gregor Mendel, were rediscovered and made known to the scientific world. Mendel had crossed varieties of garden peas that differed in characteristics such as the shape and color of the seeds or the length of the stems. He observed patterns of inheritance in the second and subsequent generations of peas, and he made sense of these patterns by recognizing that each characteristic he studied came in two versions. Today, we would call the versions of a given gene responsible for this kind of variation *alleles*. For example, a gene for seed color in Mendel's peas had two alleles—one for yellow and the other for green. When both alleles were green, the seeds in the first and second generations were green. When both alleles were yellow, the seeds in the first and second generations were yellow. But they were also yellow if one allele was yellow and the other green. That is, the yellow allele was dominant, and the green one was recessive. Today we would say that when together the yellow allele was "expressed," while the green one was not.

When an organism has two *different* alleles for a given characteristic, say, yellow and green, the organism is heterozygous for that characteristic. When an organism has two *identical* alleles for a given characteristic, we say the organism is homozygous for that characteristic. In Mendel's experiments, when two pea plants that were heterozygous for seed color were crossed, three-quarters of the offspring produced yellow seeds and one-quarter produced green seeds. The green seeds were just as green as they would be in a pure-breeding green pea plant. They were not a yellowy green. This meant that the two alleles, for green and yellow, were not affected by their coexistence in a heterozygote; they remained particulate, that is, separate.

Mendel established the important principle that there are strict rules for inheritance that can be expressed in mathematical ratios. In the case of many characteristics, if the kinds of alleles in the parents are known, the percentages of the kinds of offspring can be predicted. Furthermore, the alleles seemed to be stable—Mendel did not observe any new variations (what we would call mutations). But were these discoveries restricted to garden peas, or did they apply to other organisms? Through intense experimentation with many different species of plants, other geneticists discovered that with minor variations, Mendel's rules held for other sexually reproducing species as well.

The next major advance in understanding genetic mechanisms came in the early 1900s, when Thomas Hunt Morgan and his students at Columbia University began to study inheritance in the small fruit fly *Drosophila melanogaster.* This species is commonly found hovering above fruit, and the Morgan group collected individuals for genetic studies from ripe bananas placed on the windowsill outside their laboratory. *Drosophila* proved easy and inexpensive to raise in the laboratory, and its generation time was short—less than two weeks. This meant that the results of crosses could be determined quickly, compared with crosses of peas, mice, or human beings. It was also possible to make certain types of specific crosses with *Drosophila:* mothers to sons, or fathers to daughters, for example. Such incestuous crosses are important in detecting recessive alleles that are carried in the heterozygous state and hence are masked by the dominant alleles.

With these techniques, the Morgan group discovered many alleles for a given characteristic in fruit flies. Using these alleles as markers, they rapidly established that genes are parts of chromosomes. Previous work had established that the chromosomes, which are found in the nucleus of cells, are present in pairs—*Drosophila* has eight chromosomes in four pairs, whereas humans have 46 chromosomes in 23 pairs. Each gene is located at a fixed place in a chromosome, so in an organism every gene is represented twice—one on each chromosome of a pair. However, the gene may have one version (say, a yellow-seed allele) on one chromosome and a variant version (a green-seed allele) on the other chromosome. Morgan and his colleagues discovered that unlike the case with seed color in garden peas, most features of the fly are affected by many different genes, not just one, and a single gene usually affects several different structures. Thus, tracing the patterns of inheritance of complex structures proved more difficult than Mendelian genetics had initially led geneticists to believe.

In other experiments Morgan's group discovered that the cells that form the eggs or sperm undergo complex changes that randomly reshuffle the chromosomes with their genes. Thus, each egg or sperm gets a set of reshuffled chromosomes that is different from the chromosome of the parent and from that of every other egg or sperm. When egg chromosomes of one individual combine with the sperm chromosomes of another individual during fertilization, the result is a huge amount of variability in the offspring.

But sexual reproduction, with its associated recombination of genes, is not the only source of the variation on which natural selection can act. Another source is the formation of entirely new alleles when a gene mutates. The rate of mutation is generally very low under natural conditions, but early experiments showed that the rate can be increased by external factors such as exposure to X-rays, high temperatures, or mustard gas. The list of mutation-causing agents (mutagens) that we know today is far longer. But whereas the rate of mutation can vary depending on the stimulus chosen, Morgan and his group found that the *kinds* of new mutations were not related specifically to the stimulus—the same sorts of mutations occurred no matter what the stimulus.

This discovery of the random nature of mutation was very important for evolutionary theory. It suggested that mutations favoring a thick fur coat (in a mammal) are no more or less likely to occur in the Arctic than mutations favoring a thin fur coat. That is, mutations are not caused by the environment or by the specific needs of the organism. What causes polar bears in the Arctic to have thick fur coats rather than thin ones is natural selection acting on this random tendency to produce thicker or thinner coats. Gene mutation is constantly producing new alleles that are then screened by natural selection for their adaptability to a given environment; if they are of advantage to individuals of the species, they will eventually be incorporated into the species' genetic makeup because more individuals with the advantageous mutation will survive and reproduce than individuals without it.

When geneticists studied the alleles of individuals from populations in the wild, they discovered that these individuals were heterozygous for many of their genes. Such abundant heterozygosity, plus the reshuffling of genes when eggs and sperm cells are formed, followed by fertilization between genetically different eggs and sperm, provides for tremendous genetic variation among offspring. So much, in fact, that except for identical twins, every human being probably differs from every other human being who is now alive or who has ever lived.

Inheritance—the rules for the transmission of genes from parents to offspring—had become an extraordinarily exact science by World War II. At that point the interests of geneticists turned elsewhere: to the questions of what genes actually are and how they act. Experiments to determine the chemical structure of genes began in the 1930s, and by the 1950s James Watson, Francis Crick, and others had discovered that genes are strings of nucleotides that are part of the nucleic acid molecule deoxyribonucleic acid, or DNA. The main function of genes is to produce protein molecules.

Within a gene the four nucleotides (adenine, thymine, cytosine, and guanine, abbreviated A, T, C, and G) are combined in functional groups of triplets. Each triplet is a template for the formation of a codon of messenger RNA (mRNA), which is composed of three of the four nucleotides: U, A, G, and C. Each codon in turn determines a specific amino acid that will be added to a growing chain of amino acids. Table 2 illustrates the translation of DNA to mRNA for a short section of a gene with the code GCA GGT TAC GTC. There are 64 possible combinations of messenger RNA codons, each responsible for inserting a specific amino acid or for stopping synthesis (see table 3). This chain will eventually fold in complex ways to become a protein molecule. Many, indeed most, of the proteins that cells make are enzymes, which facilitate chemical reactions within the organism. All the biochemical events that occur within cells are the result of the activities of enzymes and other proteins. Thus, the mighty DNA with its many genes is the boss, but it is not the workforce. DNA can do only two things: it can replicate itself, which it does whenever a cell divides, and it can determine the specific proteins that cells make. It performs these functions in all organisms, from the simplest bacteria to the most complex mammals.

The messenger RNA codons are also the same in general structure

TABLE 2.

The Synthesis of a Protein Molecule

If the DNA Triplets Are	Messenger RNA Codons Will Be	Amino Acid Added to the Protein Chain Will Be
G	C	
C	G	arginine
A	U	
G	C	
G	C	proline
T	A	
T	A	
A	U	methionine
C	G	
G	C	
T	A	glutamine
C	G	

TABLE 3.

The 64 Codons of Messenger RNA

AUU	isoleucine	GUU	valine	UUU	phenylalanine	CUU	leucine
AUC	isoleucine	GUC	valine	UUC	phenylalanine	CUC	leucine
AUA	isoleucine	GUA	valine	UUA	leucine	CUA	leucine
AUG	methionine	GUG	valine	UUG	leucine	CUG	leucine
ACU	threonine	GCU	alanine	UCU	serine	CCU	proline
ACC	threonine	GCC	alanine	UCC	serine	CCC	proline
ACA	threonine	GCA	alanine	UCA	serine	CCA	proline
ACG	threonine	GCG	alanine	UCG	serine	CCG	proline
AAU	asparagine	GAU	aspartic acid	UAU	tyrosine	CAU	histidine
AAC	asparagine	GAC	aspartic acid	UAC	tyrosine	CAC	histidine
AAA	lysine	GAA	glutamic acid	UAA	(stop)	CAA	glutamine
AAG	lysine	GAG	glutamic acid	UAG	(stop)	CAG	glutamine
AGU	serine	GGU	glycine	UGU	cysteine	CGU	arginine
AGC	serine	GGC	glycine	UGC	cysteine	CGC	arginine
AGA	arginine	GGA	glycine	UGA	(stop)	CGA	arginine
AGG	arginine	GGG	glycine	UGG	tryptophan	CGG	arginine

in all species. This system is called the genetic code. Whenever a system as complex as the genetic code is found to be essentially the same in most species, we can be reasonably sure that it was established very early in the evolution of life and was conserved thereafter. Its very complexity almost ensures that it will persist, because any new system would be in competition with an already working model. There is no direct way of testing this idea in fossil organisms, but it is a satisfying naturalistic explanation for the remarkable degree to which many cell structures and cell processes have been conserved and now characterize most living organisms.

NATURAL SELECTION

The inability of Charles Darwin or anyone else in the last half of the nineteenth century to demonstrate that natural selection operates in wild populations was a serious problem for understanding evolution. Scientists at that time generally accepted that evolution had occurred, but there was no agreement that natural selection acting on variation was the mechanism, as Darwin had proposed. If natural selection was the mechanism for evolutionary change, why did one not observe species becoming better adapted to their environment? After all, animal and plant breeders could mold one variety into another in just a few generations of selective crossing. Why did nature not act with equal speed? If it did, naturalists should be able to observe evolutionary changes within their lifetime. Yet none were apparent.

A satisfactory answer to this question did not come until well into the twentieth century. First, it had to be understood that natural selection does not convert a species into the best imaginable new population; all selection can do is produce a population that continues to survive and leave offspring. Once a species has reached that level of adaptation, the pressures for becoming better adapted decrease. Natural selection is for getting by, not for producing the best possible form of life. Any natural population that survives has already withstood every assault its environment could throw at it. Once a natural population has adapted to its environment, natural selection has the important role of eliminating individuals that may be abnormal or poorly adapted because of the chance combinations of genes they inherited from their parents. Thus not only does natural selection have a creative role in providing a mechanism for adapting to new environmental conditions, but it also has a cleansing role in maintaining the population by eliminating individuals with deleterious genes. This cleaning effect is the most important role of natural selection in a population in a stable environment. If the environment changes in a significant way, however, the population is no longer as well adapted, so it has to either adapt to the new conditions or become extinct. When the environment shifts, evolutionary changes may be more rapid, and a new species may be the result.

Once these relations were understood, how one could study selection became obvious: a population adapted to one environment could be confronted with a very different environment that would challenge the population in new ways. Any changes in the population would be good evidence that natural selection was operating. Classic studies demonstrating natural selection were done by two English naturalists, J. W. Tutt in the late nineteenth century and Bernard Kettlewell a half century later. The wings of the peppered moth have a complex pale and dark spotted pattern suggested by its name. In unpolluted rural areas where the trees have light-colored lichens growing on their dark trunks, a resting moth becomes almost invisible because of its pale coloration. Lichens, however, are highly susceptible to airborne pollution, and a very different situation is found in industrialized regions. In the nineteenth century when industry began polluting the atmosphere, the lichens would die, leaving bare, dark tree trunks. A pale-colored moth on a dark trunk would be a highly visible food source for a predatory bird. It was observed, however, that not many light-colored moths inhabited industrial regions where nearly all of the individuals were very dark and hence protectively colored—a phenomenon called industrial melanism. Regions with little or no air pollution, however, still had trees with the pale, lichen-covered trunks and the pale form of the moths. Genetic experiments showed that the coloration pattern is inherited. But was there any natural selection? Careful observations consisted of watching, hour after hour, the moths on tree trunks to see which color type was caught by birds. It was found that the pale moths on dark trunks and dark moths on pale trunks were captured far more frequently than pale moths on pale trunks or dark moths on dark trunks.

In some areas where air pollution was later greatly reduced, lichens were found growing on the tree trunks once again. When this occurred, the dark form of the moth was replaced by the pale form. The observation that the evolutionary changes reverse when the environment returns to the prior condition is an important confirmation of the hypothesis that the changes are due to the levels of pollution. That is, if melanism is caused by increased pollution, a reduction in pollution should cause a decrease in melanism.

This pattern of change to the dark form and then reversal to the light form makes sense only if random mutations to the dark form are constantly occurring in the original populations of pale moths. Dark moths appearing in a rural woodland with lichen-covered tree trunks would be conspicuous to hungry birds. Natural selection in the form of those hungry birds would, therefore, eliminate the dark forms of the moths. Consequently, the moth population would remain almost entirely pale-colored. The forces of natural selection would change radically, however, when the air started to become polluted and the lichens died, leaving the dark-colored tree trunks exposed. The dark form would now hold the advantage, and the pale forms would be eliminated by natural selection. Industrial melanism is now known to occur in many species of moths from many parts of the industrial world, and it is known to decrease if pollution decreases.

Many natural populations besides moths show rapid evolutionary changes when they are confronted by a radically new environment.

When antibiotics first became available a half century ago, many previously fatal diseases were easily controlled and cured by drugs such as penicillin. This drug was widely, even excessively, prescribed, and in just a few years an alarming situation developed. Strains of bacteria that were previously destroyed by penicillin developed resistance to the drug. New types of antibiotics were then developed; but in time, bacteria mutated and became resistant to the newer drugs as well. There are similar examples of populations of insect pests, for example, house flies, cockroaches, human head lice, and the fleas of dogs and cats, becoming resistant to the chemical pesticides developed to destroy them.

The development of resistance in bacteria and insect pests are forms of evolutionary change, and it is astonishing that they happen so rapidly. We normally think of evolution as an exceedingly slow process, and this is still true in a constant environment. But when a population is confronted by new and life-threatening environmental challenges, rapid change is possible. When a population of insects encounters a highly toxic pesticide in its environment, one which it has never encountered in the past, nearly all individuals are likely to be killed. But chance mutation in a few individuals may provide a low level of resistance—enough for these few to survive and reproduce. Since the resistant individuals are the only ones remaining to breed, their genes for resistance are transmitted to the next generation. Subsequent mutations of other genes could increase resistance and so increase the percentage of survivors to the point that the genetically resistant individuals become dominant in the population. Not until resistances in bacteria were studied did it become possible to determine whether gene mutations occurred randomly or whether a specific environmental challenge caused specific mutations. Experiments showed that the frequency of gene mutation leading to penicillin resistance in bacteria is the same whether or not penicillin is present in the culture medium.

Different genes mutate at different rates. On average, a given gene might mutate once in one individual every generation in a population of a hundred thousand to a million individuals. Many populations of plants and animals are at least that large, which means that in a given generation every gene of a species is mutating at least once in some individual. Since new mutations are almost always recessive, they will be masked by the dominant allele. However, these recessive alleles are carried along in the population. The tremendous amount of reshuffling of alleles that occurs when sperm and egg cells form and in fertilization results in some individuals that are homozygous for the recessive alleles. They will provide the genetically different individuals on which natural selection can act.

RATES OF EVOLUTION

The fossil record shows that the rate of evolution varies considerably. The data are rarely good enough to provide more than estimates. Some fossil remains of mammals from the end of the Pleistocene, about 10,000 years ago, are identical with living species. The many species of mammals and birds mummified by the ancient Egyptians 3,000 to 4,000 years ago are the same as present-day individuals. These data suggest that it takes longer than 10,000 years for one species to evolve into another.

There are other ways to obtain estimates for rates of evolution. During the Pleistocene Ice Age the sea level was lowered due to the immense amount of water stored in ice sheets in the polar regions and in glaciers that covered the northern parts of the Old and New Worlds. England was then a peninsula of Europe. When the ice melted about 10,000 years ago, both England and the island of Jersey in the English Channel became isolated again. Today a few of the animals on Jersey are different enough from those on the European mainland to be classed as subspecies. Most, however, have remained the same. Other data suggest that it takes about 500,000 to 1,000,000 years for the evolution of a new species in birds and mammals.

Some general numbers can be given for the time interval from the beginning of one genus to the beginning of the next. Eight genera of fossil horses are now known that may form a lineage from *Hyracotherium* to *Equus*. *Hyracotherium* lived about 55 million years ago and *Equus* probably appeared about 1 million years ago. Therefore, if seven pre-*Equus* genera evolved in 54 million years, the average duration of each would be 7.7 million years.

The data for fossil vertebrates being among the best available, it is possible to estimate the time for the origins of the different classes of vertebrates (see figure 3, p. 84). The oldest known class, the very primitive fishlike forms called agnathans, first appeared about 500 million years ago. It took about 100 million years for some advanced agnathans to evolve into the bony fishes. After another 50 million years some advanced fishes evolved into the amphibians, and 60 million years later the first reptiles appeared. After an additional 100 million years, the birds and mammals evolved from the reptiles. Thus ancestors of the mammals spent about 50 to 100 million years in each of the lower classes. Only a few species in a given class evolved into a higher class. For instance, after the first amphibians evolved they radiated into a large number of different species. Most became extinct; others persisted and evolved as the amphibians of today—mainly frogs, toads, and salamanders—while another kind of amphibian evolved into the primitive reptiles.

The immense duration of the vertebrate classes should be compared with the relatively short period for the major evolutionary radiation of orders of higher mammals—those with a placenta. Although the first mammals evolved from reptiles in the Jurassic period, it was not until the beginning of the Tertiary period that placental mammals began to evolve their great diversity: shrews, moles, horses, cattle, whales, dolphins, anteaters, elephants, lions, wolves, bats, plus many kinds that became extinct. Nearly all living mammals are placentals, and almost their entire history has taken place within the past 65 million years.

In contrast, a few genera have shown essentially no detectable change over millions of years. For example, the horseshoe crab, *Limulus,* so common on the Atlantic coast of the United States, has remained essentially unchanged for about 185 million years, and closely similar genera are known from 360 million years ago. The lineage of a Cambrian creature similar to the living species *Peripatus*—a fascinating animal with characteristics of both the annelid worms and arthropods—suggests that this species has persisted with little change from 570 million years ago to the present.

In recent years paleontologists have emphasized that evolution does not always occur at a slow, constant rate but is characterized by long periods of little or no change followed by (geologically) short periods of rapid change. For example, the orders of mammals evolved rapidly in the early Tertiary, after which they went through a period of slow evolutionary radiation within each order. This same phenomenon may apply to the evolution of species as well; that is, change may be relatively rapid as a new species evolves, followed by a long period during which the species remains essentially unchanged. This phenomenon was named "punctuated equilibrium" by Stephen Jay Gould and Niles Eldredge. It suggests that new species arise when the environmental conditions change and natural selection begins to alter the genetic makeup of the population to adapt better to the new environmental pressures. Once these adaptive changes have been made, there is little further evolution of the new species as long as the environment remains constant.

SPECIATION

New species can evolve in two major ways. First, a parent species can evolve into a single new species. Second, a parent species can evolve into two or more new species. In the first method the number of species remains the same; in the second it increases.

If the only way to produce new species had been evolution of one species into one other species over time, there would be only one species alive today. This model is obviously not adequate to explain the huge variety of living species we see in the world around us and in the specimens of the fossil record. Clearly, evolution also occurs by a single species diverging into two or more lineages that over time become different from both the ancestral population and each other. The original gene pool becomes two independent gene pools. But how does such a split occur?

By the middle of the twentieth century a convincing hypothesis was available to answer this question. If a single population becomes geographically divided, the subpopulations become subject to evolving in different ways in response to the environmental pressures they each encounter. This pattern of evolution is called geographic speciation (see figure 11). It is not the only mechanism for speciation, but it remains the one best documented. The accumulation of the relevant data for this hypothesis started in the nineteenth century, when traveling naturalists such as Darwin and Wallace began observing differences in populations that were seemingly of the same species but geographically separated. In contrast, stay-at-home naturalists saw very little variation within the local species of animals and plants. Other naturalists observed similar cases where what appeared to be a single species was found on both sides of some barrier such as a desert or a mountain range. In each of these cases the gene pools would be at least semi-isolated, each carrying its allotment of the original species' gene pool.

By the 1930s and 1940s the evolutionist Ernst Mayr had made a convincing case for geographic speciation: The recently isolated population might find itself in a region where temperature, rainfall, vegetation, predators, or food supply differed from that of the original environment. If some of the immigrants could survive the new conditions, the forces of selection would be different and very strong—as in the case of insects first confronting pesticides. The new population's size would be much smaller than that of the original population, and as hostile environmental conditions were confronted its mortality rate would be greater. With fewer individuals to breed, more inbreed-

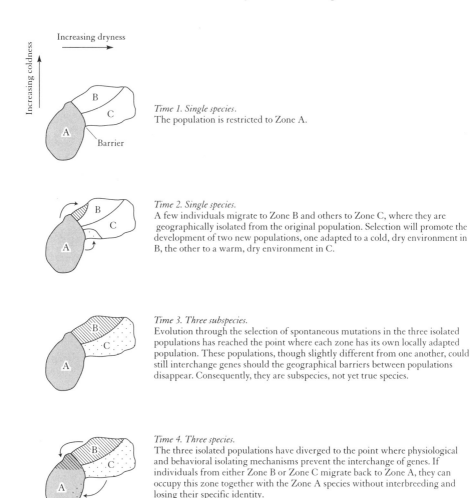

Figure 11.   A model for geographic speciation, thought to be the dominant mode for the evolution of new species. (Source: Modified from John A. Moore, *Science as a Way of Knowing,* Harvard University Press, 1993, p. 169.)

ing would occur and recessive traits would have an opportunity for greater expression. In time the newly isolated population might become so unlike the original population as to be recognized as a different species.

In the case of some species, isolation need not involve major ecological or geographical distances. Many species of parasites, for example, are restricted to a single host species, while closely similar species occur in other hosts. For instance, it is suspected that the protozoan genus *Plasmodium,* which causes malaria, was originally a generalized parasite, but in time its populations isolated specifically in birds, human beings, or other animals were each selected for improved survival in a specific host. Different host species provided the equivalent of geographic separation. After all, there are considerable barriers to cross to get from the circulatory system of a bird to that of a mammal. The spread to a new host probably occurred in the same manner by which the malaria parasite moves from one human host to another—it is carried by mosquitoes.

Evolutionary biologists are fairly comfortable with the hypothesis that evolution can occur as a consequence of usually great geographic isolation as well as microgeographic isolation on different host species. However, there is vigorous discussion but little agreement today about the possibility of other, presumably minor, mechanisms.

Once two populations become genetically different, there are other mechanisms for keeping their gene pools separated. For example, closely related species living in the same region may differ in the times of year that they breed, or they may have anatomical differences that prevent interbreeding, or they may have a behavioral preference for breeding with members of their own species or be restricted to different local habitats such as a forest or grassland. There is also the powerful mechanism of genetic incompatibility. In most cases individuals of even closely similar species cannot produce viable offspring when cross-mated. Fertilization may occur, but the embryos may die early in development. Some hybrid crosses go a step further. The embryos develop normally, but they become sterile adults. Female horses and male donkeys, for example, can breed and produce mules—prime examples of hybrid vigor. But mules do not produce normal eggs and sperm. Their set of horse chromosomes and another of donkey chromosomes differ enough from each other so that they cannot participate in the formation of normal sex cells. These various reproductive mechanisms all help to prevent interbreeding between closely related species and preserve the genetic integrity of each.

WHAT IS A SPECIES?

Consider the following: In a series of fossils belonging to the same lineage, each subsequent generation is imperceptibly different from the parent generation, but after many generations the individuals may become recognizably different in structure from the original species A. If those differences are similar in magnitude to the differences that separate living species, we would call the later population species B and we would say that species A evolved into species B. There was never a time when both species were present. Still, to be strict about defining species, we would have to arbitrarily draw a line at some generation and say that the offspring of species A are now members of species B. This is never a problem because the gaps in the fossil record eliminate the intermediate forms.

That, of course, is a ridiculous. A consensus on the definition of species might be easier to reach with respect to living species, but even in these cases, various criteria are used to separate species and no single criterion is adequate in all cases. The most useful criteria are those associated with the ability or inability of populations to interbreed, which horses and mules can do, and freely exchange genes, which they cannot do. If two populations in the wild do not exchange genes, each can maintain its genetic integrity and evolve in its own way. If they can exchange genes, their genetic integrity will be lost. It is assumed that individuals of the same sexually reproducing species belong to a single interbreeding unit; that is, any individual can in theory breed with any other individual of the opposite sex and produce viable offspring. That is a good criterion, but it cannot be used when two populations of individuals that might be capable of interbreeding are separated by a geographical barrier—such as when one population lives on the mainland and another on an offshore island. The two geographically isolated populations could not cross because there would be no opportunity for boy to meet girl. If the two isolated populations differ sufficiently from one another in ways that usually characterize different species, each could be regarded as a distinct species; if not, they would be thought of as a single species isolated into two groups.

On the other hand, the ability of two populations to interbreed under natural conditions does not mean that they are always recognized as a single species. A notable example is lions and tigers. They can be crossed in captivity, and depending on the direction of the cross, the offspring are ligers (male lion and female tiger) or tiglons (male tiger and female lion). This sorrowful miscegenation is avoided today because lions are largely restricted to Africa and tigers to Asia. Some species of wild ducks, very different in external characteristics, do interbreed but not to the degree that they cease to be different species. Thus, the individuals of a species are *usually* part of a single potentially interbreeding unit, *usually* do not interbreed with individuals of another species, are *usually* distinguishable on the basis of external characteristics and behavior patterns, and are *usually* found in a restricted geographic area where all individuals can, in theory, meet all other individuals of the same interbreeding unit.

It is fascinating to contemplate what would have been the nature of biodiversity today had extinction not been such a dominant force in the history of life—that is, if examples of every species of microorganism, plant, and animal that ever evolved remained to this day. The result would be a complete intergradation of all life, and the recognition of species would be totally arbitrary. Bacteria and elephants would be connected by populations that formed a continuum of nearly identical types. The fossil record shows, however, that every species that ever lived either evolved into another species, evolved into two or more species, or became extinct. Thus life today is not a continuum but is sequestered in distinct packages—the many millions of different species.

CLASSIFICATION

At the time Darwin was working on the *Origin* in the 1840s and 1850s, known species of plants and animals were arranged in a hierarchical classification based almost entirely on their structural characteristics. According to Darwin's theory, the similarity of species within any category of classification was due to their evolution from a common ancestor, and this is the explanation we still use today. But the current classification of living organisms, especially in poorly studied groups, is far from settled.

One basic problem is the lack of sufficient fossil data in most lineages. Another is the uncertainty surrounding selection of characteristics as the basis for classification. Most categories of classification of living species are based on their shared common characteristics and on their differences from species in other groups. Thus, the categories of classification include the likes and exclude the unlikes. This may sound vague, and it is. Classification is an art that depends on the knowledge and skill of the classifier. But the classification of species in well-studied groups, such as the flowering plants, vertebrates, and many kinds of arthropods, has reached a considerable level of stability. This means that competent systematicists—professionals whose job is to classify organisms—tend to agree on the basic classifications.

In the next few decades puzzles about classification that have persisted for generations have a good chance of being solved at last, as the new techniques of molecular biology provide quantitative data on the degrees of resemblance between two organisms. One method compares the similarities of DNA in different organisms. The long strands of DNA are extracted and broken into short segments. The DNA of two different kinds of organisms can be mixed. Since similar segments of DNA are able to combine with one another, the proportion that combines can be determined. The percentage combining is taken as an indication of relationship—the greater the percentage, the closer the relationship. When Sibley and Ahlquist (1987) used this technique of DNA-DNA hybridization to estimate the closeness of various primates, they found that human beings are much closer to the chimpanzee than to any other primate. Progressively less similar were the gorilla, orangutan, gibbon, and Old World monkeys.

Two other nucleic acids, mitochondrial DNA and ribosomal RNA, are also being used to estimate degrees of relatedness. The procedures are quite different, but they and the DNA-DNA hybridizations usually give similar results—an important affirmation of the probability that both are correct.

The ribosomal RNA data are now judged to be superior to data based solely on anatomical features. Anatomical differences are usually qualitative and subjective—the shape of a bone, for example—but molecular data can be strictly quantitative. Using ribosomal RNA molecular geneticists have discovered that the traditional order Insectivora, far from being a natural group, is a heterogeneous collection of species. The ribosomal RNA of the elephant shrew, which in the nineteenth century was assigned to the Insectivora, is very different from that of the common shrew, which is a typical insectivore. It was found to be much closer to the elephant, manatee, and hyrax (Springer et al. 1997).

The classification of organisms should soon reach a level of precision never before achieved. The data from molecular studies such as those just described will be integrated with the older data from comparative anatomy and embryology, and the existing system of classification will be adjusted accordingly. Systematics, which has been regarded by some as a backwater of biological research, has suddenly become one of the most productive and exciting fields in biology.

THE ORIGIN OF LIFE

Life had to originate before it could evolve, but we have only vague hypotheses for how this may have happened. There will probably never be direct evidence for how life began, and even if it should prove possible to synthesize life in a test tube, we would know only one way it can be done today, not how it actually happened billions of years ago. The question is so basic and interesting, however, that attempts have been made to study what the contributing events might have been.

The pattern of origin-of-life research has been to identify those special properties that distinguish living creatures from nonliving things and then to search for experimental conditions that will allow nonliving systems to exhibit some of these properties of life. A striking difference between the nonliving environment and living systems is in chemical composition. The nonliving environment—the inorganic world—consists of all of the nearly 100 natural chemical elements. The more common ones—iron, oxygen, silicon, and magnesium—together make up about 92 percent by weight of all elements in the Earth's crust. In nature the elements almost never consist of single atoms but are combined to form molecules, which may be of the same or different elements. The vast majority consist of combinations of a few elements. The oxygen in our atmosphere is a molecule composed of two oxygen atoms—O_2. The water molecule is a compound composed of two hydrogen atoms and one oxygen atom—H_2O.

Living organisms do contain some small molecules. In fact, water, the environment in which almost all chemical reactions take place, is the most abundant molecule in cells. However, the molecules that are unique to living organisms are very large, consisting of hundreds or even thousands of atoms. Only six elements—oxygen, carbon, hydrogen, nitrogen, sulfur, and phosphorus—plus a few other elements combine to form the main classes of organic compounds: proteins, carbohydrates, lipids, and nucleic acids. Compounds containing carbon are responsible for the properties that we call life.

A key feature in the origin of life, therefore, must have been a process, or processes, that converted simple inorganic compounds of the Earth's crust into complex organic compounds. Replicating such a process proved a difficult problem for chemists, and for a long time it was assumed that only living organisms could synthesize organic compounds. Then in 1828 Frederich Wöhler synthesized from inorganic chemicals an organic compound. It was urea, which is composed of only eight atoms—NH_2CONH_2.

There is no known situation in the world today, apart from living organisms who do it all the time, where complex organic compounds are being synthesized from inorganic substances. In trying to understand how life might have begun, we must search for the conditions under which complex organic compounds could be made by a nonliving system. If no such conditions exist today, could they have existed in the past? By the middle of the twentieth century, cosmologists thought that conditions during the interval between 4.8 and 3.8 billion years ago, shortly after the Earth was formed, might have been such that organic compounds formed spontaneously. That period has been well-named the Hadean eon because of its resemblance to Hades. It was a violent time of tremendous electrical storms as the molten surface of the Earth cooled to form solid rocks. For a while, showers of meteorites bombarded the Earth (Figure 8, p. 110).

Could complex organic molecules have been formed spontaneously from simpler molecules under conditions thought to prevail in the Hadean eon? In 1953 Stanley Miller performed a simple experiment to test this hypothesis. He put water (H_2O), hydrogen gas (H_2), methane (CH_4), and ammonia (NH_3) in a flask with tubes so arranged that when the water boiled, the steam containing the other molecules passed through a section of the tubing bombarded with electric sparks. The steam was then cooled to form water and returned to the flask to repeat the process.

After a week of boiling this mixture, the water had changed from colorless to colored. Something had happened. Analysis showed that amino acids, lactic acid, and other organic compounds that are the building blocks for more complex organic molecules were present. This critical experiment has been repeated many times, and all of the basic types of molecules found in living cells—amino acids, the smaller molecules from which DNA and RNA are made, and sugars—have been produced under the conditions thought to have existed on the early Earth.

In other similar experiments a higher level of complexity was reached when amino acids linked to form small proteins. Quite recently it has been discovered that RNA molecules can replicate themselves outside of a living cell. Not only that, RNA can stimulate the joining of amino acids to form proteins. These properties make RNA a candidate for being one of the first self-replicating molecules and, hence, a critical step in the origin of life.

The working hypothesis of those studying the origins of life is that the sorts of conditions that prevailed in Hadean times—and simulated in Miller's flasks—led to the oceans comprising a rich aggregation of organic compounds, a primeval soup. An important step was still required, however. Life is not a stew of oceanic dimensions; it exists in packages. The basic package is a cell. A cell consists of an integrated and interacting set of organic compounds in water, surrounded by a membrane. The membrane keeps the contents intact and prevents many external substances from entering the cell while allowing the needed molecules to enter and waste molecules to exit. The basic composition of the cell membrane is much the same in all living cells: they are composed of proteins and fatlike molecules known as phospholipids. In some very recent experiments phospholipids have been synthesized under simulated early Earth conditions, and these are able to join one another to form tiny vesicles. Furthermore, these lipid vesicles can spontaneously take in complex organic molecules, including proteins and DNA, from a surrounding solution.

These and many other experiments are exciting to those who seek to show how life *could* have originated. It is important to remember that these experiments do not show what happened at the close of the Hadean eon, but they do show that in laboratories today some basic steps in the origin of life can be simulated and studied. Even if a few decades from now it is possible to produce in the laboratory an artificial "cell" that can reproduce, this again will show only what might have happened. However, such a triumph would be satisfying to many as probably the best that can be achieved.

The implication so far is that the life of planet Earth originated on Earth. Possibly yes, possibly no. In recent years some meteorites have been found to contain organic compounds that could have been synthesized under purely physical and chemical conditions but also could be associated with life of some sort. These meteorites are thought to date to the time our solar system was forming and to be the detritus left over after the sun and planets were formed. One can even speculate that the origin of life could have involved similar steps whether on the Earth or elsewhere. The astonishing thing is that we seem to be on the threshold of developing the technology and science to answer this question.

THE ORIGIN OF MULTICELLULAR ANIMALS

The appearance of animals and plants consisting of many cells was one of the most important moments in the history of life. This event, which paleontologists now believe occurred about 1.7 billion years ago, initiated the extraordinary evolutionary radiation that populated the land and the seas with the creatures familiar to us. Before this, all life had been in the single-cell stage for the first three-quarters of its existence.

Little is known about the origin of the multicellular animals, called metazoans, and essentially all of the information has been obtained since World War II. Recently metazoans have been found in the Precambrian geological strata, but as yet they offer little information about the origin and evolutionary diversification of the animal phyla.

The first major discoveries of these Precambrian metazoans were made in the Ediacaran Hills in South Australia in the 1940s by the Australian geologist R. C. Sprigg. The strata date from about 670 million years ago—approximately 100 million years before the onset of the Cambrian. The fossils are mainly impressions of dying animals left in the mud at the bottom of a large body of water. Some paleontologists believe they are primitive jellyfish. Other remains are interpreted as the burrows of wormlike marine organisms. Subsequent to the discoveries in Australia, Ediacaran-like fossils have been found in other parts of the world. The Ediacaran organisms are so different from those appearing in the Cambrian period and later that paleontologists have yet to agree on how they should be classified. Following the period when the Ediacarans lived, the Earth passed through a time of intense cold, and this may have exterminated most metazoans then alive. The survivors were the ancestors of the species whose hard parts are found as fossils in the Cambrian strata.

A noteworthy discovery in 1909 of metazoan fossils of the middle Cambrian period was made in the Burgess Shale of Western Canada by the American paleontologist Charles D. Walcott. Many of the fossils are of soft-bodied animals, but a few have shells. Preservation of soft-bodied animals is rare at any period, so it is of great significance to have a fine sample from the time when the major animal phyla are thought to have first appeared and begun their diversification. In recent years the site and the specimens have been intensively restudied, and about 170 species are currently recognized. In contrast with the puzzles of the Ediacaran species, many of the Burgess Shale creatures can be identified as members of the major phyla that persisted and are with us today, such as the sponges, coelenterates, mollusks, annelid worms, arthropods, echinoderms, representatives of other small invertebrate phyla, and even one chordate. Primitive plants were also present. Explaining the sudden appearance of these advanced animal phyla, given their absence among the Ediacaran fauna, is a major challenge for future research.

THE ORIGINS OF HUMAN BEINGS

Darwin knew of no acceptable evidence for the antiquity of the human species. Yet it was hard to exclude human beings from the operations of evolution. Darwin almost ignored the question in the first edition of the *Origin* (1859), knowing full well how difficult it would be for such a notion to be accepted. It was not until near the end of the last chapter that he mentioned the unmentionable. In discussing how the concept of evolution could provide a rational explanation for so many of nature's puzzles, he wrote, "Light will be thrown on the origin of man and his history" (488). Many of those who understood what Darwin was implying might not have eagerly awaited the shedding of that light.

Interest in human antiquity started long before Darwin. By the middle of the nineteenth century a huge array of fossil mammals had been discovered, yet no remains of ancient human beings had been recognized as such. In fact, many doubted that any would ever be found. The French naturalist Cuvier, renowned for his knowledge of fossils, had proclaimed that there were no fossil human beings. This conclusion, perfectly valid at that time, was based on the absence of human bones in association with the bones of extinct fossil mammals. His verdict was accepted by those who adhered to the Judeo-Christian tradition. Yet there were hints that human beings may have existed. For a long time throughout Western Europe people had been aware of the presence of small hammerlike stones that seemed to have been deliberately chipped, and it was hard to believe that other than human beings could have done so. One suggestion was that they were petrified lightning bolts. Others thought they might be tools, but there was no evidence, such as fossil bones, at that time for the existence of human beings who could have made such crude tools. The evidence then available suggested that copper, bronze, and iron were the materials from which tools had always been made. The answer to this problem came when the first European explorers reached the New World in the fifteenth century. They found that the inhabitants made many of their tools from stone, not from metal. Who, then, were those early Stone Age Europeans?

In 1856, however, some strange-looking human bones were discovered under about five feet of mud at a site in the Neander Thal (valley) in Germany. They were studied and described by a skilled human anatomist, Hermann Schaaffhausen, who suggested that they were of human origin. He noted the thickness of the bones and the shape of the top of the skull, which was somewhat apelike, and suggested that the bones were of one of the ancient savage tribes that had lived in Germany. Others suggested that the bones were not old but merely deformed by some pathology. Another hypothesis was that they were the remains of a soldier in the Russian army that had marched through the area in 1814 on the way to Waterloo.

This was the first good evidence of the remains of an ancient human, now known as "Neanderthal man." Many other individual bones or parts of skeletons were found later, as were the remains of other early European human beings, the Cro-Magnons. Although both the Neanderthals and the Cro-Magnons seemed to be more ancient than 4004 B.C., their ages could not then be surmised. The Cro-Magnons were structurally like modern human beings, and the Neanderthals were only slightly less so.

It was not until the closing years of the nineteenth century that really old human remains were discovered. In 1891 a Dutch physician, Eugene Dubois, began an active search in Java for fossil remains of early human beings. A crew of convicts was assigned to help with the digging. At a depth of 40 feet they found a human tooth and the top of a skull; a year later, they uncovered a leg bone (femur). The skull was so primitive that at first Dubois identified it as a fossil chimpanzee. But the shape of the femur indicated a creature that walked upright—therefore more likely human than ape. Dubois estimated that the capacity of the cranial cavity was about 900 milliliters, compared with 1,300 milliliters for modern human beings and 400 milliliters for the chimpanzees. The fossil was given the name *Pithecanthropus erectus* ("ape man erect") to indicate its intermediacy between apes and human beings and that it walked on two feet. The name was later changed to *Homo erectus.*

This was the first discovery of a truly ancient human fossil. On the basis of later information *Homo erectus* was estimated to have lived 1.7 million to 250,000 years ago. It was assumed that *Homo erectus* evolved from a much more ancient ancestor that was also the progenitor of the modern great apes, the gorilla, orang, chimpanzee, and the bonobo. By the close of the nineteenth century most paleoanthropologists agreed that the *Homo erectus*, the Neanderthal man, and modern human beings form a sequence on the path of human evolution. But as the twentieth century was to reveal, the story is more complicated than that.

The fossils of *Homo erectus* and the Neanderthals established that individuals intermediate in structure between some unknown prehistoric apelike creatures and modern human beings had existed—as Darwinian theory demanded. This indicated that humans had evolved, but not that those two fossil types were direct ancestors of present-day human beings. Far more material would be required to establish such relationships. The search for evidence of our past history became more intense, but for several decades the discoveries were few—though more Neanderthal and Cro-Magnon remains, as well as a single jaw of another species, *Homo heidelbergensis,* were discovered in Europe.

In the 1920s important discoveries began to be made in Africa. The first, in South Africa, was of an immature individual with features intermediate between those of apes and *Homo.* It was given the name *Australopithecus* ("southern ape"). This was the first of an extraordinary group of African fossils that has thrown considerable light on our past history.

Possibly half a dozen well-defined species in the genus *Australopithecus* are now known, all from Africa. *Australopithecus* was a way station in the evolution from the very ancient apes to species of the genus *Homo.* The size of the brain in the *Australopithecus* species was 500 milliliters or less, only slightly larger than that of a chimpanzee.

One of the species of *Australopithecus* was probably ancestral to *Homo habilis,* generally regarded as the first species of *Homo* to evolve. This event occurred about 2.4 million years ago. Existing data suggest that Africa remained the hub of human evolution for thousands of years, but gradually various species migrated to Asia and Europe. Even *Homo sapiens* is thought to be African in origin, and it is the species that includes all modern human beings.

The data now show beyond all reasonable doubt that human beings have evolved over millions of years from some ancient primate that was more apelike in skeletal characteristics than are modern human beings. This means that since about 6 million years ago when the lineages leading to human beings and to the great apes diverged, ours evolved much more rapidly. Yet by some measures we are not all that different from other advanced primates. We share about 98.4 percent of our genes with the chimpanzee. We might say that the chimp is 98 percent human, or that humans are 98 percent chimp. In fact, our degree of similarity to the chimps in biological characteristics has led a prominent biologist, Jared Diamond (1992), to suggest that human beings are really a third species of chimpanzee, joining the common chimpanzee and the pygmy chimpanzee. Thus we would abolish the genus *Homo.* Nonbiological reasons will probably prevent Diamond's suggestion from being widely adopted. Alternatively, we might point out that a mere 2 percent difference in genetic makeup can have astonishing consequences.

Human evolution is a complex and rapidly advancing field of research and will remain so as long as only the estimated 3 percent of all extinct primate species have been discovered. Although information about human evolution is increasing rapidly, the data are not yet sufficient to be sure of the exact lineages. It is wiser to accept the usual caution of paleontologists and speak of intermediate or transitional forms rather than ancestors and descendants. The fossils so far uncovered form a series of transitional anatomical types, with the more ape-like ones being the oldest and those more like *Homo sapiens* being the most recent. There is another problem, so far without any generally accepted answer, and that is to understand the origin of the different types of human beings, all considered to be a single species, *Homo sapiens,* now occupying the Earth.

Thus, in spite of the important discoveries, especially in the last half of the twentieth century, a great deal still waits to be learned about evolution in general and our own in particular. But this, of course, is typical of all science: a problem solved gives us an answer but leaves us with new questions that extend and refine the target of our inquiry.

SIX

Evolution on Trial, 1925

In the nineteenth century the United States witnessed a great ferment in Christianity that resulted in many schisms that persisted into the twentieth century. The religious climate was very different from that in most nations of Western Europe, which had established or dominant religions, such as the Episcopal (Anglican) Church in England and the Roman Catholic Church in France, Italy, and Spain. In the United States, some of the sects took an extreme fundamentalist position and insisted on the inerrancy of the Bible. One of the founding fathers of American fundamentalism was the evangelist Dwight L. Moody, whose position was that a line should be drawn between the church and the world and every Christian should get both feet out of the world. Among the sins of the world to be avoided were activities on Sunday such as sports, entertainment, reading the newspaper, attending the theater, and on all days, dancing, card playing, liquor drinking, pandering to the lusts of the flesh, and atheistic teachings such as evolution (Marsden 1980).

Fundamentalists were vehemently opposed to the theory of evolution, since Genesis provided a very different account for the origin and diversity of life. In previous centuries many theologians had interpreted the Bible to say that the Earth is flat, that the Earth is the center of the universe, and that the sun rotates around the Earth. The voyage of Magellan in 1522 had provided convincing evidence that the Earth is not flat, but news seems to travel slowly, for in 1922 a schoolteacher in Kentucky was fired for teaching that the Earth is round (Ginger 1958, 63). In 1543, just a few years after Magellan's voyage, Copernicus presented evidence that the sun, not the Earth, is the center of our solar system. His views, too, were eventually accepted.

As time went on, more and more religious leaders relaxed their demands for inerrancy of the Bible relative to the discoveries that the Earth is neither flat nor at the center of the universe. The clergy were able to accept these apparent violations of Holy Writ and so could their congregations. This has not happened with evolution, however. Well into the twentieth century, William Jennings Bryan, a famous Democratic politician and even more famous orator, proclaimed that "the evolutionary hypothesis is the only thing that has seriously menaced religion since the birth of Christ and it menaces all other religions as well as the Christian religion, and civilization as well as religion" (1923, 679).

THE SCOPES TRIAL

A major challenge to the dominance of creationism and the rejection of evolution occurred in Dayton, Tennessee, in 1925. John Thomas Scopes was tried for teaching Darwinism in his high school biology class—in defiance of an act recently passed by the General Assembly of the State of Tennessee. The trial brought the evolution-creationism controversy to the nation's attention and emphasized the polar positions of the two sides.

On January 28, 1925, the lower house of the Tennessee legislature passed a bill introduced by John Washington Butler. The critical paragraph read as follows. "Section 1. Be it enacted by the General Assembly of the State of Tennessee, that it shall be unlawful for any teacher in any of the Universities, Normals and all other public schools of the State, which are supported in whole or in part by the public school funds of the State, to teach any theory that denies the story of the Divine creation of man as taught in the Bible, and to teach instead that man has descended from a lower order of animals." (Unless noted otherwise, the quotations are from the transcript of the Scopes trial, 1925.)

Butler was no Bible-thumping firebrand but a kind, friendly gentleman who had become concerned when a young woman of his community had attended a university and returned home believing in evolution but no longer in God. Butler worried that a similar fate might befall his children, and he decided to do something effective: he ran for the state legislature and was elected. Part of his platform was the need to prohibit the teaching of evolution because it might corrupt young people. He introduced a bill to do just that, and by coincidence, the well-known politician, statesman, and fundamentalist William Jennings Bryan delivered a lecture in Nashville the very night following the introduction of Butler's bill. His title was "Is the Bible True?" Bryan argued strongly that it was. Copies of his lecture were distributed to the members of the legislature and may have swayed some of them. In any case, Butler's bill passed the upper house and was signed by Governor Austin Peay, who expressed the opinion that the bill was a distinct protest against an irreligious tendency to exalt so-called science and to deny the Bible. He went on to say that the tendency was fundamentally wrong and fatally mischievous in its effects on children, institutions, and the country.

There was no organized opposition to the Butler Act, even though some politicians and others may have thought it an error to enact such a law. For office seekers it would have been folly to deny the truth of the Bible, which was the basis of the religious beliefs of most voters in Tennessee. It was the New York City–based American Civil Liberties Union that decided to test the validity of the Butler Act. The ACLU arranged for Scopes, a 24-year-old high school biology teacher and graduate from the University of Kentucky, to be charged with teaching evolution in defiance of the Butler Act. The ACLU hoped to use the trial to test the constitutionality of the law. The position of the defense was that it was not proper for any legislature to pass such a law and that prohibiting the teaching of Darwinism was analogous to prohibiting the teaching of heliocentrism—that the sun is the center of the universe. Clarence Darrow, a prominent trial lawyer, became the chief lawyer for the defense. Bryan offered his services to the prosecution, saying that the state could not afford to have a system of education that destroys the religious faith of children.

The trial began in Dayton on Friday, July 10, 1925. The small town's population was greatly swollen by numerous visitors from out of state, more than a hundred newspaper reporters, local people who came in from the surrounding hills and farms, and merchants to supply the needs of all for food, trinkets, toy monkeys (representing ancestors), and Bibles. The Scopes trial was seen as big news. It defined the antagonists and the concerns about teaching evolution that are still with us at the beginning of the third millennium. The trial was a contrived event initiated by the ACLU and supported by those modernists who sought to prevent religious points of view from deciding what was to be taught in the public schools. On the opposing side, the defenders of the Butler Act were determined to maintain their way of life and system of belief.

Presiding Judge John T. Raulston called the court to order. He read the Butler Act and then the first chapter of Genesis. Interestingly enough, the Bible was identified as the "St. James version rather than the King James version" (66). The jury was instructed not to pass judgment on the constitutionality of the Butler Act but simply to decide whether or not Scopes was guilty of teaching evolution. The main lawyer for the prosecution, Attorney General A. T. Stewart, read that section of the state constitution relative to freedom of religion, which said that no preference should be given to any religious establishment or mode of worship. Darrow, the defense lawyer, suggested that the Butler Act gave clear preference to the Bible and asked why not, in

that case, allow the Koran as well. The attorney general answered that
the Bible was preferred because they were not living in a heathen
country.

On the second day of the trial Darrow made a lengthy statement
indicating what the defense hoped to accomplish, namely, to have the
Butler Act declared unconstitutional. Darrow held Bryan guilty for
such a pernicious law's being passed in the first place: he had a long
history of antievolution activity and had encouraged, without success,
the passage of an antievolution law in Florida. This excerpt and others
that follow will give the flavor of Darrow's remarks.

> I remember, long ago, Mr. Bancroft [a historian] wrote this sen-
> tence, which is true: "That it is all right to preserve freedom in
> constitutions, but when the spirit of freedom has fled, from the
> hearts of the people, then its matter is easily sacrificed under law."
> And so it is, unless there is left enough of the spirit of freedom in
> the state of Tennessee, and in the United States, there is not a sin-
> gle line in any constitution that can withstand bigotry and igno-
> rance when it seeks to destroy the rights of the individual; and
> bigotry and ignorance are ever active. Here we find today as bra-
> zen and as bold an attempt to destroy learning as was ever made
> in the middle ages, and the only difference is we have not pro-
> vided that they shall be burned at the stake, but there is time for
> that. (75)

Darrow then argued that the law was invalid because it was not
drawn up properly. Further, it was wrong to specify the Bible as *the*
divine book and ignore the Koran, the Book of Mormon, the beliefs
of Confucius or Buddha, or even Emerson's essays on transcendental-
ism. Darrow was aware of the "higher criticism" of the Bible because
he informed the court that there are two conflicting accounts of creation in the first two chapters of Genesis. The Butler Act had not said
which account of creation was at issue. About why Scopes was being
charged Darrow said, he knew he was there because the fundamen-

talists were after everybody who thought, because ignorance and big-
otry were rampant. Darrow ended his speech on this second day of
the trial as follows:

> I will tell you what is going to happen, and I do not pretend to be
> a prophet, but I do not need to be a prophet to know. Your honor
> knows the fires that have been lighted in America to kindle reli-
> gious bigotry and hate. You can take judicial notice of them if
> you cannot of anything else. You know that there is no suspicion
> which possesses the minds of men like bigotry and ignorance and
> hatred. . . .
>
> If today you can take a thing like evolution and make it a
> crime to teach it in the public school, tomorrow you can make it a
> crime to teach it in the private schools, and the next year you can
> make it a crime to teach it to the hustings or in the church. At
> the next session you may ban books and the newspapers. Soon
> you may set Catholic against Protestant and Protestant against
> Protestant, and try to foist your own religion upon the minds of
> men. If you can do one you can do the other. Ignorance and
> fanaticism is ever busy and needs feeding. Always it is feeding
> and gloating for more. Today it is the public school teachers, to-
> morrow the private. The next day the preachers and the lecturers,
> the magazines, the books, the newspapers. After a while, your
> honor, it is the setting of man against man and creed against
> creed until with flying banners and beating drums we are march-
> ing backward to the glorious ages of the sixteenth century when
> bigots lighted fagots to burn the men who dared to bring any in-
> telligence and enlightenment and culture to the human mind. (87)

Much of the third day was given over to a discussion of whether
or not each session of the court should begin with a prayer, and the
judge decided that having prayers would be a good thing. In a minor
episode, the attorney general characterized one of the defense lawyers
as "the agnostic counsel for the defense" and told another, Mr. Hays,
to shut up. The next day the attorney general offered his apology, and

Mr. Hays responded: "Permit me to say personally that there are two qualities I much admire in a man. One is that he is human and the other is that he is courteous. The outburst on yesterday proves that the attorney-general was human, and the apology proves that he has the courtesy of a southern gentleman" (97).

On the fourth day Judge Raulston read a long opinion on why the motion of the defense to quash the indictment was rejected. In the afternoon the attention of the trial finally turned to Scopes. The position of the prosecution was simple: there was a law that forbade the teaching of evolution, and Scopes had done so. Even the defense agreed to that.

The position of the defense was much more complicated. It held that it was necessary not only to prove that Scopes had taught evolution but also that he had denied the theory of creation given in the Bible. But the Bible has more than one theory of creation (the P and J versions), so which one had Scopes denied? Further, the defense maintained that there is no conflict between evolution and Christianity and noted that in earlier years Bryan had held the liberal view that religion was not subject to legislation and that no science can be taught without recognizing evolution. The defense asked that the fundamentalist Bryan of today revert to the modernist Bryan of yesterday. It pointed out that Christianity had survived in spite of all the discoveries of science and that science occupied a field of learning separate and apart from that of theology. The defense mentioned vestigial organs as indicative of evolution, but no evidence was presented for the evolution of human beings from monkeys, as the prosecution had said. Further, the defense argued that the Bible cannot be regarded as an adequate treatment of science because of the many discoveries made since it was written. For example, "Moses never heard about steam, electricity, the telegraph, the telephone, the radio, the aeroplane, farming machinery, and Moses knew nothing about scientific thought and principles from which these vast accomplishments of the inventive genius of mankind have been produced" (116).

Witnesses for the prosecution were then called to testify that Scopes had admitted teaching the section on evolution that was in the text-book approved by the state for use in its schools—George William Hunter's *Civic Biology*. Scopes said he had done so because he thought that the statute was unconstitutional. Surprisingly, the jury was ex-cluded when Darrow and the other defense lawyers called witnesses for the defense. The first and only expert witness permitted to testify was the well-known zoologist Maynard M. Metcalf, who went over the scientific evidence for evolution. Judge Raulston did not permit the other scientists or any of the theologians present to testify for the defense. However, they were allowed to submit written statements, which were to be printed in the official record of the trial.

When Judge Raulston asked Darrow if he wished to call Scopes to the witness stand, Darrow replied that every charge against the de-fendant was true, and when Darrow was asked again, he replied that there was no point in doing so. Scopes never was called as a witness. Again, the strategy of the defense was to have the Butler Act declared unconstitutional, not to have Scopes found innocent. If he was found guilty, the case could be appealed to the Supreme Court of Tennessee and the Butler Act would be evaluated and, it was hoped, declared unconstitutional.

A high point was in the afternoon of the fifth day when Bryan made his first serious comments. He had come to Dayton as a nation-ally known orator and politician, having served three times as the Democratic Party's candidate for president. Impressive in appearance, he was a leader of the fundamentalists and a vigorous and effective antievolutionist. The people of Dayton regarded Bryan with love, re-spect, and reverence. They expected him to demolish the forces of evil that were attempting to corrupt their children (presumably adults were immune to this infection).

Bryan began by ridiculing evolution, and the court record shows that he was interrupted frequently by laughter from those attending the trial. He referred to a diagram in the biology textbook used by

Scopes and his students showing the animal kingdom and the number of species in each major taxonomic group. Each group was in a circle with a size relative to the number of species. The species numbers were estimates, of course, but Bryan noted that all were round numbers and wondered if animals bred that way. The 3,500 mammals rated only a very small circle with hardly enough room for human beings. "There is the book! They were teaching your children that man was a mammal and so indistinguishable among the mammals that they leave him there with thirty-four hundred and ninety-nine other mammals (laughter and applause)" (175). Bryan also read from Darwin's *Descent of Man* that man evolved from Old World monkeys: "Not even the American monkeys but from Old World monkeys (laughter)" (176). "Talk about putting Daniel in the lion's den? How dare these scientists put man in a little ring like that with lions and tigers and everything that is bad! Not only the evolution is possible, but the scientists possibly think of shutting man up in a little circle like that with all these animals that have an odor, that extends beyond the circumference of this circle, my friends (extended laughter)" (175).

Since the defense had been prohibited from having all but one scientist testify about evolution, they asked about having expert witnesses for the Bible. Byran replied: "Now, your honor, when it comes to Bible experts, do they think they can bring them in here to instruct the members of the jury, eleven of whom are members of the church? I submit that of the eleven members of the jury, more of the jurors are experts on what the Bible is than any Bible expert who does not subscribe to the true spiritual influences or spiritual discernments of what our Bible says (Voice in audience, 'Amen!')" (180–81).

Bryan also sought to show that some distinguished scientists had doubts about evolution. He quoted from a lecture given in Toronto by the prominent British geneticist William Bateson to prove this point. The defense countered this point with a statement by one of the scientists who had submitted in writing what he had intended to say. That scientist had written to Bateson to ask his views. Bateson replied that he had looked through his Toronto address and found nothing that could be construed as expressing doubt regarding the main fact of evolution. He then expressed the opinion that the campaign against the teaching of evolution was a terrible example of the way in which truth could be perverted by ignorant people.

Dudley Field Malone, one of the defense lawyers, answered Bryan at length. He pointed out the difference between theological and scientific ways of thinking: "The main difference between the theological mind and the scientific mind is that the theological mind is closed, because that is what is revealed and is settled. But the scientist says no, the Bible is the book of revealed religion, with rules of conduct, and with aspirations—that is the Bible. The scientist says, take the Bible as guide, as an inspiration, as a set of philosophies and preachments" (184). Malone continued:

> There is never a duel with the truth. The truth always wins and we are not afraid of it. The truth is no coward. The truth does not need the law. The truth does not need the forces of government. The truth does not need Mr. Bryan. The truth is imperishable, eternal and immortal and needs no human agency to support it. We are ready to tell the truth as we understand it. . . . We feel we stand with progress. We feel we stand with science. We feel we stand with intelligence. We feel we stand with fundamental freedom in America. . . . We ask your honor to admit the evidence as a matter of correct law, as a matter of sound procedure and as a matter of justice to the defense in this case (profound and continued applause). (187–88)

Judge Raulston was not swayed, and the attorney general claimed that evolution would be the end of the Bible:

> I say, bar the door and do not allow science to enter. That would deprive us of all the hope we have in the future to come. And I say it without any bitterness. I am not trying to say it in the spirit of bitterness to a man over there [Darrow], it is in my

view, I am sincere about it. Mr. Darrow says he is an agnostic. He is the greatest criminal lawyer in America today. His courtesy is noticeable—his ability is known—and it is a shame, in my mind, in the sight of a great God, that a mentality like his has strayed so from the natural goal it should follow—great God, the good that a man of his ability could have done if he had aligned himself with the forces of right instead of aligning himself with that which strikes its fangs at the very bosom of Christianity.

Yes, discard that theory [Divine Creation] of the Bible—throw it away, and let scientific development progress beyond man's origin. And the next thing you know, there will be a legal battle staged within the corners of this state, that challenges even permitting anyone to believe that Jesus Christ was divinely born— that Jesus Christ was born of a virgin—challenge that, and the next step will be a battle staged denying the right to teach that there was a resurrection, until finally that precious book and its glorious teachings upon which this civilization has been built will be taken from us. (197–98)

On Friday, the sixth day of the trial, Judge Raulston read his long ruling explaining why he intended to exclude the scientists from giving testimony before the jury. This caused Darrow to make some extremely critical and sarcastic remarks to the court, which could be considered in contempt. The judge was offended, and Darrow was required to post a bond for $5,000 while the judge decided what to do about Darrow's remarks. The trial day was short; the court adjourned at 10:30 A.M. The weekend must have been tense for all concerned. On Monday Darrow offered an apology and Judge Raulston responded: "My friends, and Col. Darrow, the Man that I believe came into the world to save man from sin, the Man that died on the cross that man might be redeemed, taught that it was godly to forgive and were it not for the forgiving nature of Himself I would fear for man. The Savior died on the cross pleading with God for the men who crucified Him. I believe in that Christ. I believe in these principles. I accept Col. Darrow's apology" (226).

Among other items of business on Monday, a letter from the governor of Tennessee was read. It stated in part: "After careful examination I can find nothing of consequence in the books now being taught in our schools with which this bill will interfere in the slightest manner. Therefore, it will not put our teachers in any jeopardy. Probably the law will never be applied" (214). The governor seemed to be making a strong point for the defense, but Judge Raulston ignored the letter and ruled that the courts, not the executive branch of government, would decided what the laws meant. "His opinion of what the law means, whether or not it would be enforced, is of no consequence at all in the court, and could not have any bearing, and I exclude the statement" (214). The defense was not permitted to refer to the state's newly adopted biology textbook, which presented evolution as an extraordinarily important concept in biology. This was the day that the statements for the defense prepared by the scientists and religious leaders were read into the record.

Later in the day, one of the most astonishing events in the trial occurred. After a lengthy discussion of the many variants of the Bible, the defense wished to question witnesses on the Bible, and Bryan agreed to take the stand as an expert. The jury missed all of this. Darrow asked the questions and Bryan answered. Here is a sampling (284–304):

DARROW: Should everything in the Bible be literally interpreted?

BRYAN: Yes, everything literally except some illustratively such as man being the salt of the earth.

DARROW: Was Jonah literally swallowed by a whale?

BRYAN: I think it was a big fish.

DARROW: Did Joshua make the sun stand still?

BRYAN: I believe what the Bible says.

DARROW:	Did the Flood of Noah occur and, if so, when?
BRYAN:	Yes.
DARROW:	Do you know that the date of 4004 B.C. was determined from the generations given in the Bible?
BRYAN:	I am not sure.
DARROW:	What do you think about the date?
BRYAN:	I do not think about things I don't think about.
DARROW:	Do you think about things you do think about?
BRYAN:	Sometimes.
DARROW:	Do you think the earth was made in six days?
BRYAN:	Not six days of twenty-four hours. [This answer was greeted with gasps from the fundamentalists in the courtroom—they were literalists and if the Bible said days, that meant a day consisting of 24 hours.]
DARROW:	Was the first woman Eve?
BRYAN:	Yes.
DARROW:	Was she made from Adam's rib?
BRYAN:	Yes.
DARROW:	Where did Cain get his wife?
BYRAN:	I do not know.

Each question was explored in detail, and slowly Bryan began to crumble. He proved to be a poor witness when it came to questions about the Bible, and he knew he was letting down those in the courtroom who believed in the inerrancy of biblical statements. The attorney general interrupted to ask the purpose of Darrow's line of questioning—hoping to put an end to it:

BRYAN:	The purpose is to cast ridicule on everybody who believes in the Bible, and I am perfectly willing that the world shall know that these gentlemen have no other purpose than ridiculing every Christian who believes in the Bible.

DARROW: We have the purpose of preventing bigots and igno-
           ramuses from controlling the education of the United
           States and you know it, and that is all.

BRYAN: I am simply trying to protect the word of God against
           the greatest atheist or agnostic in the United States
           (prolonged applause). I want the papers to know I am
           not afraid to get on the stand in front of him and let
           him do his worst. I want the world to know (prolonged
           applause) (299).

On Tuesday, July 21, 1925, the eighth and last day Darrow re-
minded the court that Scopes had taught evolution as charged, including that humans had evolved from a lower order of animals, so
the jury should be instructed to find Scopes guilty. The jury was
brought back into the courtroom, and Judge Raulston gave his instruc-
tions. The jury retired, deliberated for nine minutes, and found Scopes
guilty. The judge set the fine at $100—an amount that was to prove
critical as events unfolded. The court asked Scopes if he had anything
to say. He did. "Your Honor. I feel that I have been convicted of
violating an unjust statute. I will continue in the future, as I have in
the past, to oppose this law in any way I can. Any other action would
be in violation of my ideal of academic freedom—that is, to teach the
truth as guaranteed in our constitution, of personal and religious freedom. I think the fine is unjust" (313).

So the case came to its close. Bryan had this to say:

> Causes stir the world. . . . Here has been fought out a little cause of
> little consequence as a case, but the world is interested because it
> raises an issue, and that issue will some day be settled right,
> whether it is settled on our side or the other side. It is going to be
> settled right. There can be no settlement of a great cause without
> discussion, and people will not discuss a cause until their attention
> is drawn to it, and the value of this trial is not in any incident of the
> trial. . . . [T]his case will stimulate investigation and investigation

will bring out information, and the facts will be known, and upon the facts, as ascertained, the decision will be rendered. . . . [N]o matter what our views may be, we ought not only to desire, but pray, that that which is right will prevail, whether it be our way or somebody else's. (316–17)

Darrow was not so philosophical: "I think this case will be remembered because it is the first case of this sort since we stopped trying people in America for witchcraft because here we have done our best to turn back the tide that has sought to force itself upon this—upon this modern world, of testing every fact in science by a religious dictum. That is all I care to say" (317).

Darrow had argued for what was right. Bryan had argued for what was righteous. Darrow had lost. He had hoped to show that the Butler Act was unconstitutional because it violated freedom of speech and religion and to call distinguished scientists and theologians for their testimony of what the theory of evolution and the Bible said about creation. Judge Raulston overruled all of these requests. It was admitted that Scopes had taught evolution, as charged; so unless the Butler Act could be shown to be unconstitutional, there could be no other verdict than guilty.

All in all, the Scopes trial was a bizarre affair. The textbook that contained the discussion of evolution had been selected years before by the State of Tennessee for use in its schools. Apparently this choice had never upset parents, whether they were fundamentalists or not. The stimulus for the trial was not local discontent but the desire of the ACLU in New York for a test case. A citizen of Dayton had to convince Scopes to say that he had taught evolution and so to be tried; Scopes was never asked what had really happened. Had he been asked, there would have been no case, because after the trial ended he confessed to a newspaper reporter that he had not been in school on the day evolution was discussed in his biology class. Another teacher had substituted for him!

Bryan had prepared a long speech to give before the jury, but he never had the opportunity. After Darrow finished his devastating questioning of Bryan about the Bible, the attorney general thought Darrow might savage Bryan once again and so decided it would not be wise to allow Bryan to give his speech even to redeem himself. Byran did give copies of the undelivered speech to the press after the trial ended. He spent the next few days traveling and making speeches. On Sunday, July 26, five days after the trial ended, Bryan was back in Dayton. After a hearty dinner, he died in his sleep. Some say this was in part a consequence of the deep humiliation he had suffered at the trial. Others thought his death was a consequence of overeating. Maybe both are true.

In a final attempt to have the Butler Act declared unconstitutional, the verdict in the Scopes trial was appealed to the Supreme Court of Tennessee, which was embarrassed by the whole affair and had a difficult time deciding what to do. Happily for the justices, they found a way to avoid a decision. The jury in the Scopes trial had returned a verdict of guilty, but they did not specify the fine. The amount was left to Judge Raulston, and he set it at $100. That was a big mistake. The constitution of Tennessee required that fines of more than $50 be determined by the jury. Therefore the Supreme Court declared a mistrial and reversed the decision that Scopes was guilty. The usual next step would have been to hold a new trial. However, the Supreme Court felt that nothing was to be gained by prolonging the life of this "bizarre case" and suggested that the attorney general forget the entire matter—which he did.

AFTER DAYTON

The Scopes trial did nothing to resolve the debate between the creationists and the evolutionists. The lawyers for the defense were unable to have the act declared unconstitutional either at the trial or upon appeal to the Supreme Court of Tennessee. The mistrial meant that the prosecuting lawyers had failed as well, because Scopes was not convicted. The Butler Act remained part of the law of Tennessee, but it was rarely invoked and was finally repealed in 1967. Other states, mainly in the southeast, considered bills banning the teaching of evolution. In most cases they did not pass. In Kentucky, for example, an antievolution bill was introduced, but it was laughed away—another bill introduced at the same time demanded that water should run uphill. Some church leaders and church groups even came out in support of teaching evolution.

On balance, however, the Scopes trial proved to be a plus for the creationist movement. Bryan, as a chief spokesman and revered leader of the nation's creationists, gave national stature and brilliant oratory to the cause. Not unexpectedly, his followers were shocked and dismayed when under Darrow's searing cross-examination Bryan proved to be inept, especially in suggesting that the "days" of creation might not have been of 24-hour duration but possibly were very much longer. Bryan overcame this lapse from orthodoxy by conveniently dying less than a week after the trial ended. Darrow was blamed for contributing to this outcome, and Bryan became a martyr—a powerful plus for any religious cause.

The ascendancy of creationism had a chilling effect in the classroom. The solution for many teachers was to omit evolution altogether. Teaching in public schools is difficult enough, and it is not made easier when parents complain that their children's sacred beliefs are being undermined and that they are "turning from God." Most teachers in the 1920s were not well prepared to teach evolution anyway—a situation that persists to this day. Textbooks reduced their coverage of evolution and placed any discussion of it at the very end of the book. This placement allowed a teacher to say that "we never had time to get to that topic." Rarely was evolution presented as the grand organizing theory that makes so many biological things and processes understandable.

A notable example of the effect of the Scopes trial can be seen in the history of the textbook *Biology for Beginners* by Truman J. Moon, which was the predominant high school biology textbook in the nation for decades. In the 1921 edition, published four years before the Scopes trial, the frontispiece carried a full-page portrait of Charles Darwin, and a short chapter titled "The Development of Man" began as follows:

> With an egotism which is entirely unwarranted, we are accustomed to speak of "man and animals" whereas we ought to say "man and *other* animals," for certainly man is an animal just as truly as the beasts of the field. . . .
>
> As soon as man became intelligent enough to make comparisons between himself and other animals, the resemblances became apparent and led to the idea that some relationship must exist with lower forms. Two thousand years ago the Greeks discussed this fact and advanced various theories to account for it.
>
> Very gradually, information accumulated, and the idea of relationships developed into the theory that not only man but all other living things, both plant and animal, are not only related, but actually descended from common ancestors. This is called the theory of descent, or *evolution*. (321)

Moon then listed the evidence for evolution—such as rudimentary organs, embryological resemblances, homologous organs, geological data, domestication of animals and plants—and followed this with a family tree showing the evolutionary relations of human beings and the great apes. There was no discussion of the forces that Darwin thought were responsible for evolution—genetic variation and natural selection. After the Scopes trial, the picture of Darwin was removed from the front of the 1928 edition of Moon's textbook, and the 1933 edition omitted evolution entirely.

The biology actually taught from the late 1920s to the 1960s consisted mainly of detailed descriptions of the structure of organisms "from amoeba to man." That approach not only was far easier for the teacher to implement but also avoided unpleasant confrontations with parents about evolution. Although professional biologists at that time might have desired an adequate treatment of evolution in textbooks, decisions about content were made primarily by publishers, whose major concern was meeting the needs of the largest sector of their market. Consequently they conveniently avoided controversial subjects.

A movement in the late 1950s throughout the United States to improve science education, especially in high schools, was stimulated by Russia's launching of the satellite *Sputnik* in 1957. American leaders saw that accomplishment as indicative of better science education in the Soviet Union. Separate national committees for physics, mathematics, chemistry, and biology were formed to study the problem of improving science instruction, and each included university scientists, high school teachers, and school administrators. The biologists formed the Biological Sciences Curriculum Study (BSCS) and, with support and encouragement from the National Science Foundation, prepared experimental textbooks intended for use in tenth-grade biology courses.

The BSCS Steering Committee proposed to get away from the "parade-of-the-animal-and-plant-kingdoms" approach, to stress concepts and experimental science, and to encourage the personal involvement of students in their learning—especially in the laboratory. The two areas in biology that had been prominently ignored in high school biology courses were human reproduction and evolutionary biology. Both were to be adequately treated in the new BSCS books. One of the members of the BSCS Steering Committee, the distinguished geneticist and Nobel laureate H. J. Muller, proclaimed that "a hundred years without Darwin are enough."

The BSCS produced three textbooks, each emphasizing a different approach to biology but all including evolution and human reproduction. These controversial topics could be included because the contracts made with the publishers contained a clause giving full control of content to the BSCS. Some publishers were not happy with this arrangement, yet the extensive publicity surrounding BSCS and the other national curriculum projects suggested large sales for the new kind of books. Evolution became a major theme in BSCS biology books in the 1960s and, having been given the imprimatur of what was seen as a national reform effort, it spread to other biology textbooks as well. For a few years it seemed as though the "hundred years without Darwin" were over. Not quite.

SEVEN

The Rise of "Creation Science," 1963

The reappearance of evolution in biology courses proved to be a stimulus for creationists, and their voices and activities increased. At first, only a few creationists were actively involved outside their community, but they proved to be skillful, determined, and effective. For example, the efforts largely of one couple in Texas were sufficient to make the adoption of biology books that discussed evolution extremely difficult in that state.

The major voices for creationism were those of ten men with advanced university degrees who in 1963 formed the Creation Research Society and later, in 1972, founded the Institute for Creation Research, an educational institution with faculty, students, and research programs. A pamphlet from the Creation Research Society—issued at the society's own creation in 1963—gave a brief history of the movement, a list of activities, and requirements for membership. The credo of the organization read:

1. The Bible is the written Word of God, and because it is inspired throughout, all its assertions are historically and scientifically true in all the original autographs. To the student of nature, this means that the account of origins in Genesis is a factual presentation of simple historical truths.

2. All the basic types of living things, including man, were made by direct creative acts of God during the creation week described in Genesis. Whatever biological changes have occurred since creation week have accomplished changes only within the original created kinds.

3. The great flood described in Genesis, commonly referred to as the Noachian flood, was a historic event worldwide in its extent and effect.

Two members of the institute, Henry M. Morris (now retired) and Duane Gish, were until recently the principal professional creationists in the United States. They, along with other members of the institute, set the pattern for the activities of the antievolutionists since the 1960s by promoting what they claim is an alternative to the scientific evidence for evolution, namely, creation science. Creation scientists refuse to accept merely on faith that the story of creation in Genesis is true and to ignore the scientific evidence for evolution. Instead, they try to discredit the scientific evidence for evolution and to assemble their own body of scientific evidence to support the P version of creation in Genesis.

THE CREATION SCIENTISTS' APPROACH

The "bible" of the creation science movement is *Scientific Creationism,* edited by Morris and first published in 1974. It appeared in two editions, a general edition and one for public schools. The general edition discusses the evidence used by evolutionists and suggests that none of it proves that evolution occurred. The concluding chapter of the general edition attempts to present data that prove divine creation and are in full accord with a literal interpretation of the Bible. The public school edition "deals with all the important aspects of the creationist-evolution question from a strictly scientific point of view, attempting to evaluate the physical evidence from the relevant scientific fields without reference to the Bible or other religious literature. It demonstrates that the real evidences dealing with origins and ancient history support creationism rather than evolutionism" (iv).

The major ideas developed in creation science deny the general conclusions of biologists and geologists. For example, the Earth is assumed to be very young, 10,000 years being the outside limit. The authors reject radiometric dating by arguing that no one was present to see when the strata of the geological column were laid down. Since there can be no *direct* evidence of age, any such estimates must be uncertain at best. We can be certain that no evolutionist observed events about 4.5 billion years ago, but come to think of it, was there any creationist on hand 10,000 years ago to record what happened? Many things in science, especially historical subjects such as evolution, can be studied only by seeking indirect evidence of past events—the detective's approach. The utility of radiometric dating has long been accepted by scientists, especially when two different methods give essentially identical results. There was no way that the formation of ancient strata could have been seen by geologists.

An important implication of evolution is that there must have been intermediates, or transitional forms, between major groups if members of one group evolved into another. Many examples of intermediates are now known, and thus many "missing links" are in fact no longer missing. *Scientific Creationism* argues these new discoveries away with a surprising analysis. It notes that *Archaeopteryx* has both feathers and teeth and that feathers are characteristic of birds and teeth of reptiles. One paleontologist is quoted who classifies *Archaeopteryx* as a bird because it has feathers. The creationists conclude, "Thus, *Archaeopteryx* is a bird, not a reptile-bird transition. It is an extinct bird that has teeth. Most birds don't have teeth, but there is no reason why the Creator could not have created some birds with teeth" (85). This is a fascinating solution, but it is not based on an understanding of how the biological system of classification works. The system of classification has no place for intermediates. *Archaeopteryx* could also have been classified as a reptile—as some fossils of it actually were. Creationists could have argued just as easily that *Archaeopteryx* was an extinct reptile with feathers and that there is no reason why the Creator could not have created some reptiles with feathers.

An important feature of creationism is the belief that the Noachian flood was a worldwide phenomenon that submerged the highest mountains for a period of months. All life was destroyed except the few individuals of every species that survived with Noah and his family in the Ark. This flood was responsible for depositing vast amounts of silt that became the layers of sedimentary rocks, with their fossil organisms. The creationists' date for this cataclysm is about 2350 B.C. (247).

Another creationist, Austin (1994), has proposed that the Grand Canyon of the Colorado River was formed a few thousand years ago during Noah's flood. The data and arguments he uses are not those accepted by professional geologists, and a detailed response to Austin's theory has been provided by Elders (1998). Yet the familiar problem remains. To a reader with little or no knowledge of geology and of what counts as evidence in science, Austin seems to make a strong case. Professional geologists, however, agree that there is no geological evidence whatsoever for a worldwide flood that covered the highest mountains and destroyed all life less than 5,000 years ago.

Scientific Creationism ends with the following statement:

> There seems to be no possible way to avoid the conclusions that, if the Bible and Christianity are true at all, the geological ages must be rejected altogether. In their place, as the proper means of understanding earth history as recorded in the fossil-bearing sedimentary rocks of the earth's crust, the great worldwide Flood so clearly described in Scripture must be accepted as the basic mechanism. The detailed correlation of the intricate geophysical structure of the earth with the true Biblical framework of history will, no doubt, require a tremendous amount of research and study by Bible-believing scientists. Nevertheless, this research is urgently

needed today in view of the world's increasing opposition to the Biblical Christian faith. The vast complex of godless movements spawned by the pervasive and powerful system of evolutionary uniformitarianism can only be turned back if their foundation can be destroyed, and this requires the re-establishment of special creation, on a Biblical and scientific basis, as the true foundation of knowledge and practice in every field. This therefore must be the primary emphasis in Christian schools, in Christian churches and in all kinds of institutions everywhere. It is hoped that this book will provide the information necessary to undergird and energize the movement. (255)

Scientific Creationism seems to have done little, in fact, to "undergird and energize" the "Bible-believing scientists" to discover the data that will throw out what biologists and paleontologists have learned during the last few centuries about Earth's history. Expeditions to Mount Ararat have found there something that suggests the remains of Noah's Ark, but the evidence is not conclusive. Other creationists visit sites where geologists have found that the forces of mountain building have not only pushed up layers of rocks but turned them over. Geologists rely on study of the fossils in the strata to explain what happened, namely, that the strata have been reversed and, hence, the oldest are on the top, not the bottom. Creationists use these upside-down strata to argue that the whole story of successive layers, each representing a capsule of time, is wrong.

One of the most interesting attempts to refute the notion of the immensity of geological time and the unique fossils in the strata is the creationists' claim that human footprints and dinosaur footprints occur in the same geological strata. This would mean, of course, that dinosaurs and human beings coexisted—a notion that is definitely not compatible with the history of life as understood by biologists and paleontologists. Current estimates place the extinction of dinosaurs at about 65 million years before the appearance of human beings. Those controversial footprints are in the Cretaceous rocks along the banks of the Paluxy River in Texas. Some have been identified by geologists as indeed caused by dinosaurs. Other depressions are called human footprints by some local people including religious leaders. These have been examined extensively by geologists, and it is now accepted that the "human footprints" have a variety of origins. Some are badly eroded dinosaur footprints. Others have been carved by local citizens and hence are the result of human hands rather than human feet. Some local people recount that in the 1930s stoneworkers made a business of carving dinosaur and human footprints in the same slab and selling them to devout visitors who wished clear evidence for the correctness of Genesis and the denial of Darwin. After many years of debate and careful study by geologists, the better-informed creationists have admitted that there are no human footprints in the Cretaceous strata along the banks of the Paluxy. This myth continues, however, in the minds of many rank-and-file creationists.

Another creationist argument against evolution is that it violates the second law of thermodynamics. This fundamental law of the physical sciences holds that matter achieves and maintains complexity only with a constant supply of energy. If animals do not have a constant input of energy from food, they die and their complex structure decays to relatively simple molecules. Green plants are able to use the energy of sunlight for the synthesis of their living substance. Animals use the products of the plant world to produce their own structures and the energy to maintain those structures. If an animal was in a closed system, that is, with no supply of external energy, the existing energy in the molecules that comprise its body would slowly be lost and death ensure. Some creationists argue that since evolution entails an increase in the complexity of organisms—from simple one-celled creatures to whales, elephants, redwood trees, and human beings—it violates the second law. Thus there can be no evolution from simplicity to complexity. Creationists seem to forget that they, along with nearly all complex animals, began life as a single cell, which developed into a complex adult—an example of a level of complexity the second law fails to prevent. Why? This is possible because organisms do not live in a closed environment—food is required to sustain lives.

Evolution per se does not require additional energy—it involves only individuals being born, growing up, reproducing, and dying. Organisms use energy whether or not they are evolving. They evolve only if the proportions of different genetic types in the offspring differ from those in the parental generations. I suspect that not one of the prominent creationists believes his argument about the second law of thermodynamics, but reference to it can have a powerful effect on a naive audience. So if it works, why not use it?

The creationists have not been able to present any evidence supporting their arguments for a young Earth, a worldwide flood, or an interpretation of the paleontological data that would support the accuracy of the P version in Genesis. They have, however, been very successful in presenting their case to the public. In recent years Gish and other creationists have lectured widely to church groups and to college and university communities both in the United States and abroad. Until the 1990s, the lectures followed a standard pattern and dealt with just a few topics. The footprints on the Paluxy were presented repeatedly as was the second law of thermodynamics. Both in lectures and in published works such as *Scientific Creationism* acceptance of the creationists' arguments depends on an inadequate understanding of science or firm religious beliefs.

Scientific Creationism is well-written and is far less demanding than a scientific account of evolution. For a poorly informed person it would probably be more appealing than a scientific book on evolution. The creationists' story is easy to understand and may support religious beliefs that have been held since childhood. The evolutionists' story demands a more open mind and a good background in science—or else a simple faith that scientists are presenting their findings in a fair and honest manner.

Professional creationists spend a great deal of their time on speaking tours, and some of them are quite up-to-date on recent research in evolutionary biology and paleontology. Thus they know where the gaps are in the data for evolution. The goal of the creationists is to emphasize the gaps and sow doubt in the minds of the audience about the adequacy of the evidence for evolution. They might pose a question like: "Evolutionists claim that whales evolved from land-living mammals. Where is the evidence?" Until recently there was none. The scientists' answer is that there is considerable evidence from the fossil record that traces the evolution of some species, but one cannot expect to have fossil evidence for the origins of all living creatures. Most fossil records are and probably always will be incomplete. A few well-documented cases are sufficient, and this number increases yearly. For example, just a few years ago fossils that indicate an intermediate stage between a terrestrial mammal and a whale were discovered in Pakistan.

Using this skeptical approach, a skilled creationist can easily best an evolutionist in debate. The audience can be made to wonder why there is no evidence to answer the creationist's specific question. A poorly informed audience is not always prepared to accept that if there are a few cases showing the evolution of a group of species—the horses, for example—that is enough. In a sense this is a matter of the public's not understanding how science works and what is accepted as proof. No science is ever complete; its statements represent only the best analysis at the time, based on the best available evidence. Consider the case of medicine: Although the medical profession knows a great deal about many diseases, everything is not known about any one disease. The procedures of medicine represent the best that can be done with what is known. Nevertheless, a sick person is unlikely to refuse the services of a physician by maintaining, "I will not accept medical treatment for heart disease until the profession knows enough to cure the common cold."

Picking away at the gaps in evolution is a reasonable and effective approach if the purpose is to destroy that point of view and imply that the alternative, divine creation, must be correct by default. The goal of such a strategy is, most certainly, not to provide an alternative rational explanation for the diversity of living animals and plants. Quite often the biologist or paleontologist who debates a creationist fails to realize that the creationist's purpose is not to inform but to win the battle for a particular religious point of view. The episode is more like a court trial, where each lawyer's goal is to win for the person he or she represents. If the victory for the client is also a victory for justice, so much the better, but justice is not the lawyer's primary concern.

In spite of the careful way creationists might present their case, in the final analysis it is based on belief in the story of divine creation told in the P version of Genesis, not on scientific evidence. The creationists' approach is to claim that evolution cannot be correct because there are too many unknowns, and therefore to conclude that the Genesis story must be true by default. But even if evolution could be falsified, why should the default position be Genesis? What about the many different accounts of creation that are parts of the sacred traditions of other religions? What is the basis for excluding them? Creationists have no logical answer for that question.

Most individuals who have informed themselves deeply about the data agree that evolution provides a much better account for the changes in the history of life over time and for the diversity of life we see today than does creation science, which is nothing more than old-fashioned creationism—without any science. The name *creation science* is regarded by many noncreationists as an oxymoron because creation implies a supernatural force, and science can never use supernatural forces to explain its data.

Biologists and paleontologists have assembled information fully compatible with the concept of evolution as guided by natural processes, whereas none of the few types of data that creationists offer in support of divine creation pass muster in the sciences. Much remains to be discovered about evolution, but those who study these problems and are involved in new discoveries accept the overarching concept of evolution as true beyond all reasonable doubt. Most antievolutionists around today are people who are unfamiliar with these data and with the procedures used to gain scientific understanding, or who choose to reject the theory of evolution for religious or other reasons.

THE NEW CREATIONISM

In the 1990s, the style and substance of the professional creationists' attacks on evolution changed somewhat in response to the times. The leaders of the movement downplayed or failed to accept the standard antievolution arguments such as the supposed incompatibility of evolution with the second law of thermodynamics, the supposed lack of a convincing number of fossils that are intermediate between major taxonomic groups of organisms, and the reputed footprints of human beings alongside those of dinosaurs in the geological strata along the banks of the Palusky River. The errors of these arguments came to be accepted by many creationists in the last decade.

The cutting edge of the creationist movement is now moving beyond the traditional literal interpretation of Genesis. The position that creation as described in the P version of Genesis is right and so evolution must be wrong is now being replaced by the argument from design. This notion is that some aspects of life and living species are so complicated and complex that they cannot be explained by science. There must have been an Intelligent Designer to have been able to create something as complex as the human eye, for example.

The intelligent design movement is a slight variant on the position of the natural theologians of the nineteenth century and before. It was one that appealed to many naturalists and others such as Charles Darwin and Thomas Henry Huxley. One of its most prominent advocates was the theologian William Paley, who in 1802 argued that some of the structures of animals are so beautifully and effectively adapted to their environment that they could not have "just happened." There must have been a designer, namely, God. Paley's analysis, appearing when little was yet known about living organisms, made a powerful impression in England in the nineteenth century. It was an argument more acceptable to many devout Christians than a belief that creation was accomplished in only six days and as recently as 4000 B.C.

This is Paley's famous opening of his *Natural Theology:*

> In crossing a heath, suppose I pitched my foot against a *stone,* and were asked how the stone came to be there; I might possibly answer, that, for any thing I knew to the contrary, it had lain there for ever: nor would it perhaps be very easy to show the absurdity of this answer. But suppose I had found a *watch* upon the ground, and it should be inquired how the watch happened to be in that place; I should hardly think of the answer which I had before given, that, for any thing I knew, the watch might have always been there. Yet why should not this answer serve for the watch as well as for the stone? Why is it not as admissible in the second case, as in the first? For this reason, and for no other, viz. that, when we come to inspect the watch, we perceive (what we could not discover in the stone) that its several parts are framed and put together for a purpose, *e.g.* that they are so formed and adjusted as to produce motion, and that motion so regulated as to point out the hour of the day. (Paley 1811, 1–2)

Paley described many of the anatomical and other characteristics of living organisms and argued that just as there is a watchmaker who makes a watch for the purpose of telling time there must have been the equivalent of a divine watchmaker for designing the wonderful structures of the species of living creatures. That was, of course, God. Paley concluded:

> I shall not, I believe, be contradicted when I say, that, if one train of thinking be more desirable than another, it is that which regards the phenomena of nature with a constant reference to a supreme intelligent Author. . . . The world thenceforth becomes a temple, and life itself one continued act of adoration. The change is no less than this, that, whereas formerly God was seldom in our thoughts, we can now scarcely look upon any thing without per-

ceiving its relation to him. Every organized natural body, in the provisions which it contains for its sustentation and propagation, testifies a care on the part of the Creator, especially directed to these purposes.... The works of nature want only to be contemplated. When contemplated, they have every thing in them which can astonish by their greatness: for, of the vast scale of operation, through which our discoveries carry us, at one end we see an intelligent Power arranging planetary systems, fixing, for instance, the trajectory of *Saturn,* or constructing a ring of two hundred thousand miles diameter, to surround his body, and be suspended like a magnificent arch over the heads of his inhabitants; and, at the other, bending a hooked tooth, concerned and providing an appropriate mechanism, for the clasping and reclasping of the filaments of the feather of the humming-bird. We have proof, not only of both these works proceeding from an intelligent agent, but of their proceeding from the same agent.... Therefore one mind hath planned, or at least hath prescribed, a general plan for all these productions. One Being has been concerned in all.

Under this stupendous Being we live. Our happiness, our existence, is in his hands. All we expect must come from him. Nor ought we to feel our situation insecure. In every nature, and in every portion of nature, which we can descry, we find attention bestowed upon even the minutest parts. The hinges in the wings of an *earwig,* and the joints of its antennae, are as highly wrought, as if the Creator had nothing else to finish. (539–42)

Thus, a watch fulfills its purpose because the watchmaker designed it to do so, and the species of animals and plants live their specialized lives because of the ways the Supreme Author designed them. At the time Paley was writing, little was known of the details of how our bodies function. It seemed impossible to understand how a structure as complex as the human eye might function or to accept that it "just happened."

Of course, no evolutionist today would claim that eyes "just happen." The working hypothesis is that they are the consequence of the selection of gene mutations that initially produced small areas in the skin that were light-sensitive in very simple creatures. Slowly, over great expanses of time and in a succession of species, the light-sensitive areas are thought to have become larger and more complex. After many millions of years they had become the vertebrate eyes. At all stages, it is assumed mutations were selected that made the light-sensitive areas more effective. In other words, being able to detect light was highly adaptive for some types of animals.

There is no sequence of fossils that shows this proposed evolution of the vertebrate eye. The soft parts of organisms are almost never fossilized, and in situations such as this the best that can be done is to search for living animals with light-sensitive structures of varying levels of complexity. It is now known that there exists a sequence of living animals, with a graded series of ever more complex light-receiving structures, that shows what the evolutionary story might have been. The story is far from complete, but one no longer need invoke supernatural processes to explain the eye—as Paley was forced to do.

Natural theology provided an alternative to biblical theology in explaining the natural world. There was still a causal agent, the God of Genesis or the Intelligent Designer of the natural theologians. God remained in His firmament with the natural theologians seeking to know Him by studying nature and the theologians seeking knowledge through biblical analysis. Both groups continued to rely on a supernatural force. The change in thought patterns that began in the Enlightenment made an increasing number of individuals restive with explanations of natural phenomena that invoked the supernatural. Studies of phenomena in astronomy, physics, and chemistry were increasingly able to reject the supernatural and to provide explanations that involved only natural identifiable things and processes.

In the middle of the nineteenth century Charles Darwin was able to add biology to those other sciences able to ignore supernatural explanations. Nevertheless creationists have continued to exist, and the intelligent design advocates are now active and effective with the general public, effective in part because the scientific community tends to ignore them completely.

The intelligent design program is rejected by scientists and others because I.D. creationists believe that some things about organisms are so complex that they are not just unknown but ultimately unknowable and, hence, are the work of an Intelligent Designer. This point of view was perfectly valid when there was essentially no knowledge of the physiology of cells on which life depends. How, for example, could one ever understand what must be an incredibly complex series of chemical reactions that occur with great speed in the cells of animals and plants that were so small as to be almost invisible? In the years since World War II, with the aid of a battery of new techniques, the enormous complexity of cell metabolism has been documented. Today there seems to be no aspect of the life of a cell that is unknowable, a situation similar to geologists gaining the ability to determine geological time. At the end of the nineteenth century it seemed that there could never be a method that would give confirmable answers. The discovery of radioactivity eventually provided a series of acceptable methods for geological clocks that tell accurate time and, most important, agreed with one another as to what the time is.

Fundamental to the commitment of any working scientist is the position that what we do not know today we may know tomorrow. The advance of science so far has shown this approach to be highly successful, and scientific knowledge continues to expand at an ever-increasing rate. Perhaps some things will never be known, especially events in the distance past for which no direct or indirect evidence exists. But even if we never know, for example, the mating habits of dinosaurs, that lack of concrete knowledge does not add up to an argument that dinosaurs did not mate and evolve. Recent books by Phillip Johnson (1991, 1997) and Michael Behe (1996) give more information about the intelligent design position. A balanced analysis is provided by Pennock (1999).

What can be the explanation for this continuing argument of crea-tionists and evolutionists? Two social scientists, Francis B. Harrold and Raymond A. Eve (1995), have studied the creationist movement for years and have this to say:

> Average Americans are ill-equipped to evaluate the claims of sci-entific creationism. They can hardly be expected to analyze crea-tionist claims about the second law of thermodynamics, for in-stance, if they lack any idea of what the first or third laws are about. If they have no notion of the wealth and range of evidence for human evolution, they may find perfectly reasonable the claim that all *Homo erectus* fossils are either apes or modern humans. If they subscribe to the common misconception that evolutionary change is the result of "blind chance," they may perceive creation-ist arguments of its extreme improbability as compelling. If they have little understanding of the roles of critical thinking and "rules of evidence" in science, they may cling to unsubstantiated beliefs even after seeing them effectively refuted; even if they do change their minds, they may revert after a time to their former beliefs. The efforts of scientific creationists have not affected the scientific consensus on evolution because creation scientists do not function in the research tradition of modern science. . . . They launch scattershot attacks on the evolutionary consensus based on the notion that evidence that damages evolution is evidence that supports creationism. Their actual criticisms of scientific finds are characterized by misunderstandings, omissions, and distortions. They do not attempt to develop detailed alternative scientific ex-planations of the data they claim to have "unexplained" in their criticisms of the consensus. Instead, they appeal to the supernatu-ral. . . . It seems safe to conclude that creationist rhetoric has in-deed bolstered the ideology's credibility . . . [Their] primary inter-est is not so much to develop an alternative science as to defend a traditional worldview or even advocate the establishment of a conservative Christian theocracy. In either of these cases, there would be no role left for science to play, except when its findings

could be used to support scripture. Modern science, generally, which rationally deduces hypotheses from existing knowledge, tests them with data, and then lets the philosophical and theological chips fall where they may, would have little place in a society such as the creationists envision. (91–92)

BACK TO THE COURTS

From 1923 to 1928, Arkansas, Mississippi, Oklahoma, and Tennessee passed laws forbidding the teaching of evolution. In other states such efforts were defeated. In 1968 the Supreme Court of the United States, in *Epperson* v. *Arkansas,* struck down the Arkansas statute on the grounds that it was in conflict with the establishment clause of the First Amendment of the Constitution of the United States, which reads as follows: "Congress shall make no law respecting an establishment of religion." The court interpreted creationism as based on religion and as promoting the views of a particular sect—the Judeo-Christian tradition.

Having failed in the courts to ban evolution from the classroom, the creationists tried a different approach. If it was impossible to keep evolution out of the schools, then an attempt should be made to get creationism in. They argued that evolution was "just a theory," using the term *theory* to signify an iffy hypothesis. After all, no scientist ever claimed to have seen one species evolving into another, so creationists argued that evolution could hardly have been considered a fact. Creationism as derived from the P version in Genesis could not be proven either, so both were just different ways of looking at nature. A just solution, therefore, would be to present both views and let the students decide. This was the equal-time position.

Scientists indeed cannot object to creationists saying that evolution is "just a theory," because scientists themselves use the word *theory* quite readily to describe evolution and other major natural phenomena, such as the germ theory of disease, the theory of gravitation, cell theory, or genetic theory. In common usage, the word *theory* usually means a questionable idea—a hypothesis as yet unproven. This is the standard way the word is used in creationists' speeches and in much of public discourse. It is also the way legislatures and school boards use the term when they seek to banish evolution from the schools. But scientists use the word *theory* to mean a large, overarching concept that organizes a vast body of data.

Wendell Bird, a lawyer then on the staff of the Institute for Creation Research, prepared a draft statement advocating equal time for the "theories" of evolution and of creation as described in Genesis. This draft was used by Paul Ellwanger, a therapist and then head of a creationist group called Citizens for Fairness in Education, to prepare a model bill for state legislatures to consider. It was the basis of bills introduced in about two dozen state legislatures. Only in Arkansas and Louisiana were bills passed, and both were challenged in the courts.

So for the second time in just a few years Arkansas became a focal point in the evolution-versus-creationism debate. Arkansas' law, called the Balanced Treatment for Creation-Science and Evolution-Science Act (Act 590), was signed by the governor on March 19, 1981. It demanded that equal time be given to the teaching of evolution and creationism. On May 27 of that year a suit, *McLean* v. *Arkansas Board of Education,* was filed challenging the constitutional validity of Act 590 on the grounds that it violated the establishment clause of the Constitution. Judge William R. Overton, of the Arkansas Supreme Court, heard the case.

The plaintiffs, who wished Act 590 to be declared unconstitutional, assembled an impressive group of scientists to discuss the nature of science and of creation science. Among them were Stephen Jay Gould of Harvard University and William V. Mayer, Director of the Biological Sciences Curriculum Study. There were also professional philosophers and religious leaders among the witnesses for the plaintiffs. The defendants—those favoring equal time—were unable to produce any well-respected scientists to support their arguments. They never called upon either Morris or Gish of the Institute for Creation Research, out of fear that the court would probably realize that their "creation science" was not based on scientific evidence at all. This was the best opportunity yet for the creationists to present their case, but they could develop no more than a feeble position.

Judge Overton rendered his judgment on January 5, 1982, and concluded with the following analysis:

> In the 1960's and early 1970's, several Fundamentalist organizations were formed to promote the idea that the Book of Genesis was supported by scientific data. The terms "creation science" and "scientific creationism" have been adopted by these Fundamentalists as descriptive of their study of creation and the origins of man. . . . The fact that creation science is inspired by the Book of Genesis and that Section 4(a) [of Act 590] is consistent with a literal interpretation of Genesis leave no doubt that a major effect of the Act is the advancement of particular religious beliefs. . . . The conclusion that creation science has no scientific merit or educational value as science has legal significance in light of the Court's previous conclusion that creation science has, as one major effect, the advancement of religion. The essential characteristics of science are: (1) It is guided by natural law; (2) It has to be explanatory by reference to natural law; (3) It is testable against the empirical world; (4) Its conclusions are tentative, i.e., are not necessarily the final word; and (5) It is falsifiable. Creation science . . . fails to meet these essential characteristics. Implementation of Act 590 will have serious and untoward consequences for students, particularly those planning to attend college. Evolution is the cornerstone of modern biology, and many courses in public schools contain subject matter relating to such varied topics as the age of the earth, geology, and relationships among living things. Any student who is deprived of instruction as to the prevailing scientific thought on these topics will be denied a significant part

of science education. Such a deprivation through the high school level would undoubtedly have an impact upon the quality of education in the State's colleges and universities, especially including the pre-professional and professional programs in the health sciences. (942)

Judge Overton had agreed with the plaintiffs that Act 590 was unconstitutional. This was an astonishing victory for the teaching of evolution, and it had the effect of abolishing efforts in other states to teach creationism along with evolution, for a few years at least.

In 1987 the United States Supreme Court, in *Edwards* v. *Aguillard,* struck down a Louisiana act that required equal time for the teaching of evolution and creationism. Once again, it was judged that the basis for the law specifying that creationism be taught was religious belief, not science, so the law was in conflict with the establishment clause. The majority opinion of the court was written by Justice William Brennan. One of his key statements was:

> The preeminent purpose of the Louisiana legislature was clearly
> to advance the religious viewpoint that a supernatural being cre-
> ated human kind. The term "creation science" was defined as em-
> bracing this particular religious doctrine by those responsible for
> the passage of the Creationism Act. . . . The legislative history
> therefore reveals that the term "creation science" as contemplated
> by the legislature that adopted this Act, embodies the religious be-
> lief that a supernatural creator was responsible for the creation of
> human kind (United States Reports 1990, 591–92)

and, hence, this is prohibited by the establishment clause.

The following year, in 1988, the attorney general of Tennessee was asked if a teacher in a public school was free to teach all theories of the origin of life for the purpose of enhancing the effectiveness of science instruction. His answer was no, if the theory had a religious basis. This conclusion was based on various legal opinions, especially the then-recent *Edwards* v. *Aguillard* case in Louisiana. The attorney general made it clear, however, that biblical accounts of creation could be included in courses in history, ethics, or comparative religion.

Neither Judge Overton's ruling in the 1982 Arkansas case, nor that of the United States Supreme Court in the 1987 Louisiana case, nor any of a handful of others—all dismissing the teaching of creationism as science—ended the creationists' efforts to curb or mitigate the teaching of evolution. The contest shifted from state legislatures to departments of education, school boards (Kansas being a notable example in 1999), and individual schools. Among the many creationist-inspired events that have taken place recently are the introduction of antievolution bills in state legislatures (none has passed), actions taken by local school boards to restrict or abolish the teaching of evolution, textbooks failing to win adoption because they give too much attention to evolution, the active teaching of creationism in many schools in spite of the decisions that this is in conflict with the Constitution, the requirement of disclaimers in textbooks such as "evolution is just a theory," state science curriculum standards omitting evolution, and an especially innovative step—the gluing together of the offending pages discussing evolution in the textbooks used by students. This last strategy may in fact guarantee that teenagers will become tremendously eager to learn everything they can about evolution—or anything else that adults glue shut.

Legal opinions—all the way up to the United States Supreme Court—prohibiting the teaching of creationism have not brought peace to the classrooms. The most destructive effect of the evolution-versus-creationism controversy remains the chilling effect it has had on biology teachers. Although science has prevailed in major legal challenges to the point that there are currently no laws on the books in any state prohibiting the teaching of evolution or permitting equal time for creationism, the problems for men and women trying to teach evolution are probably as difficult in some communities today as they have ever been.

EVOLUTIONISTS FIGHT BACK

The increased activity of professional creationists and the spread of their views beginning in the 1960s brought forth a major reaction from religious, educational, and scientific organizations. Scientists reacted promptly, and statements against the teaching of creationism or against giving it equal time with evolution were made by the National Academy of Sciences, American Association for the Advancement of Science, American Anthropological Association, Association of Physical Anthropologists, American Astronomical Society, American Chemical Society, American Geophysical Union, American Institute of Biological Sciences, American Physical Society, American Psychological Association, American Society of Biological Chemists, American Society of Parasitologists, Geological Society of America, Society for the Study of Evolution, Society of Vertebrate Society, and many state academies of science and organizations of science teachers.

A typical statement was made in 1972 by the Commission on Science Education of the American Association for the Advancement of Science, the world's largest scientific society:

> The Commission on Science Education of the American Association for the Advancement of Science is vigorously opposed to attempts by some boards of education and other groups to require that religious accounts of creation be taught in science classes.
>
> During the past century and a half, the earth's crust and the fossils preserved in it have been intensively studied by geologists and paleontologists. Biologists have intensively studied the origin, structure, physiology, and genetics of living organisms. The conclusions of these studies are that the living species of animals and plants have evolved from different species that lived in the past. The scientists involved in these studies have built up the body of knowledge known as the biological theory of the origin and evolution of life. There is no currently acceptable alternative scientific theory to explain the phenomena.

The various accounts of creation that are part of the religious heritage of many people are not scientific statements or theories. They are statements that one may choose to believe, but if he does, this is a matter of faith, because such statements are not subject to study or verification by the procedures of science. A scientific statement must be capable of test by observation and experiment. It is acceptable only if, after repeated testing, it is found to account satisfactorily for the phenomena to which it is applied.

Thus the statements about creation that are part of many religions have no place in the domain of science and should not be regarded as reasonable alternatives to scientific explanations for the origin and evolution of life. (Matsumura 1995, 26)

The prestigious National Academy of Sciences issued in 1984 a lengthy and authoritative statement, *Science and Creationism: A View from the National Academy of Sciences,* in support of evolution and in opposition to the teaching of creationism. A second edition was released early in 1999. The academy has also produced a book especially designed for schoolteachers, *Teaching about Evolution and the Nature of Science* (1998).

Many religious groups have also issued statements against the teaching of creationism in schools. The General Assembly of the United Presbyterian Church in the United States developed this position in 1982 from which the following is extracted:

Affirms that, despite efforts to establish "creationism" or "creation-science" as a valid science, it is teaching based upon a particular religious dogma as agreed by the court *(McLean v. Arkansas Board of Education).*

Affirms that, the imposition of a fundamentalist viewpoint about the interpretation of Biblical literature—where every word is taken with uniform literalness and becomes an absolute authority on all matters, whether moral, religious, political, historical or scientific—is in conflict with the perspective on Biblical interpretation characteristically maintained by Biblical scholars and theological schools in the mainstream of Protestantism, Roman Catholicism and Judaism. Such scholars find that the scientific theory of evolution does not conflict with the interpretations of the origins of life found in Biblical literature.

Affirms that, exposure to the Genesis account is best sought through the teaching about religion, history, social studies and literature, provinces other than the discipline of natural science, and

Calls upon Presbyterians, and upon legislators and school board members, to resist all efforts to establish any requirement upon teachers and schools to teach "creationism" or "creation science." (Matsumura 1995, 107–8)

Among the other religious groups that have issued similar resolutions opposed to the teaching of creationism are the American Jewish Congress, the Central Conference of American Rabbis, the General Convention of the Episcopal Church, the Lexington Alliance of Religious Leaders, the Lutheran World Federation, the Roman Catholic Church, the Unitarian Universalist Association, and the United Methodist Church. (The full texts are provided in Matsumura 1995.)

Not unexpectedly, many civil liberties organizations such as the American Civil Liberties Union, which was deeply involved in the Scopes trial, have issued statements. Part of a long ACLU position paper reads as follows:

Among the problems "creation science" creates in the academic environment is the foreclosure of scientific inquiry. The unifying principle of "creationism" is not the law of nature, but divinity. A divine explanation of natural data is not subject to experiment, it cannot be proved untrue, it cannot be disputed by any human means. Creationism necessarily rests on the unobservable; it can exist only in the ambiance of faith. Faith—belief that does not rest on logic or on evidence—has no role in scientific inquiry.

Vigilance requires firm and consistent opposition to every effort to use the nation's schools to teach any biblical text, including Genesis, as literal truth, either directly or disguised as "alternative" science. To reject creationism as science is to defend the most basic principles of academic integrity and religious liberty. (Matsumura 1995, 159–60)

The position of the Catholic Church is of special interest when we remember its centuries-long opposition to scientific discoveries that provided evidence contrary to the official Church position. The Catholic Church has accepted evolution, the most definitive statement being made by Pope John Paul II to the Pontifical Academy of Sciences on October 22, 1996 (for complete text and evaluations, see John Paul II 1997). The pope said that the theory of evolution is more than a hypothesis and that "it is indeed remarkable that this theory has been progressively accepted by researchers, following a series of discoveries in various fields of knowledge. The convergence, neither sought nor fabricated, of the results of work that was conducted independently is in itself a significant argument in favor of this theory" (382). The question of the human soul must be looked upon in a slightly different manner, and John Paul II agrees with his predecessor, Pius XII, who in the encyclical *Humani generis* held that "if the human body takes its origin from pre-existent living matter, the spiritual soul is immediately created by God" (383). John Paul then pointed out that although physical continuity—that is, the evolution of human beings—can be studied, the moment that the soul enters the body cannot. This assertion is not a problem for evolutionists because the soul cannot be identified or studied by the methods of science; it remains in the realm of religious beliefs. John Paul's statement is of great importance because it permits Roman Catholics to accept evolution. Nearly half of the world's two billion Christians are Catholics, and in the United States they are by far the largest denomination, totaling about 60 million.

Thus, as the twenty-first century begins there is a consensus among scientists, biblical scholars, most theologians of the mainstream religions, and educators that the theory of evolution provides a verifiable account for the origins of life and its changes over time, and that acceptance of evolution does not require a denial of one's religious beliefs.

EIGHT

Where Does This Leave Us?

Western civilization has developed two major ways of accounting for the diverse species of creatures alive today. The oldest, and probably the one most widely accepted in the United States, is based on a literal interpretation of Genesis, the first book in the Judeo-Christian Bible, which dates to a few centuries B.C. This is the familiar story of God's creation of the universe, our solar system with its sun and orbiting planets, and living creatures—all in six days. Strict creationists date these events to no more than 10,000 years ago. According to this account, a second dramatic influence on living creatures occurred when essentially all life, except for one or a few pairs of each kind or species, was destroyed by a worldwide flood. The few survivors were those taken aboard the Ark by Noah. All present-day life, therefore, is descended from the passengers on the Ark. The date of the flood is estimated to be in the third millennium B.C., possibly 2350 B.C. Creationists believe that all species have remained essentially the same as when they were created.

The entire evidence for the creationists' explanation for life on earth is found in Genesis. There is no independent confirmation of this account of creation from biology or geology, and, in fact, a tremendous body of information from these sciences argues that the creationists' stories cannot be correct. Creationists accept divine creation solely on the basis of their belief in a supernatural God, combined with an unshakable faith in the inerrancy of the Bible.

The second way of explaining the diversity of life is evolution. The modern version of this concept began its vigorous growth with Charles Darwin in 1859 and has expanded during the past century and a half. It is so important in our understanding of life that it has been said that nothing in biology makes sense without it. In contrast to the creationists' explanations that involve the supernatural, Darwin proposed that the concept of evolution must be based entirely on natural phenomena. An exclusion of supernatural things and processes is basic to the scientific approach that has given us the astonishing products of technology, modern medicine, a much more productive agricultural system, and, indeed, modern civilization.

And for the inquiring mind, we now have an understanding of the natural world that is, itself, a thing of beauty. The seeming chaos of our world and much of the cosmos is being replaced by an understanding based on a few overarching principles. One of the more basic of these is the absolute dependence of all life on the sun. Life is a consequence of the interactions of molecules in thousands of different chemical reactions. Some of these reactions release energy, but the net result is that energy from outside an organism is required for the chemical processes of life, which consist of the synthesis of simple molecules mainly into complex proteins, fats, carbohydrates, and nucleic acids. The ultimate source of this energy is the sun, but animals cannot bask in the sun and acquire the energy to keep their metabolic activities going. Most rely on green plants, which capture the radiant energy of the sun to combine simple molecules such as carbon dioxide, water, and a few salts into simple sugars. Species of the animal world are absolutely dependent on plants for food and or on other animals that eat plants. This food supplies the substance both for reassembling the plant molecules into animal molecules and for the energy necessary to do so.

The sun has other roles essential for life. Its radiant heat keeps much of the world at temperatures that permit water to be in the liquid state. This is critical since our bodies consist mainly of water, and the reactions in our cells occur in a watery environment. Heat from the sun evaporates the water that later comes back to earth as rain so necessary on the land masses that would otherwise be lifeless deserts. Rain also shapes much of the Earth's surface since it is one of the main factors in the erosion of mountains to plains and the carving of the surface into valleys by rivers and streams. Neither life nor the surface of the earth as we know it could exist without liquid water.

Vast quantities of the sun's energy from long ago are stored in deposits of coal, petroleum, and natural gas that can be used directly or indirectly in the generation of electricity for our technology, transportation, and heat and light. These natural resources, finite and precious, are in a sense fossil sunbeams. In so many ways the sun is the dynamo that drives our Earth. Its powerful role could not have been fully understood in ancient times, but, nevertheless, many early cultures worshipped the sun, often as their most important deity.

Once life appeared on Earth, that it would evolve was almost axiomatic. The simple reason is that the central property of life is its ability to reproduce, so an ever-increasing amount of life requires an ever-increasing quantity of resources. Apart from sunlight, all the resources required for life are finite, including a place to live, which is the most limiting resource for both plants and animals. The continents and the oceans may vary in size over geological time, but the sum cannot increase. The chemical substances that make up living bodies were always available not because they were unlimited but because they were cycled. Green plants take in carbon dioxide from the air along with salts and water from the soil and combine them to synthesize the chemicals that form their structures and that supply them with the chemical energy required for their life. Animals, by contrast, require oxygen from the air, water, and complex organic molecules in their food, which originate in green plants. The molecules involved enter organisms and are constantly returned to the environment as waste products and when the organism—plant or animal—dies. Molecules are borrowed for a life but not destroyed.

By Darwin's time, it was understood that the resources of the environment are limited: a single acre could not support a stand of trees, a herd of cattle, or a flock of sheep of infinite size. In 1798 Thomas Robert Malthus (1766–1834), an English economist and demographer, had suggested that human beings faced similar restrictions: populations of infinite size could not be supported by a finite environment. Charles Darwin read Malthus's work carefully and claimed that it provided the clue for how evolution might occur. It was also common knowledge in Darwin's day that a single plant in its lifetime produces a vast number of seeds, yet, by and large, the number of flowers in a field or trees in a forest remains fairly constant from year to year. Animals also have the ability to produce far more offspring than there are parents; yet the population size of any species seems to fluctuate around a mean. There was little evidence to suggest that species of animals and plants somehow restrict their own reproduction in order not to exceed the carrying capacity of the environment. Quite the contrary: many more offspring are produced than the environment can support, and nearly all the offspring of plants or animals die before reaching maturity—the world is simply not big enough for everyone.

Could there be qualitative differences between the few that survive and the many that do not? Darwin speculated that there might be. It was common knowledge that the offspring produced by two parents can differ in their characteristics. If some of these characteristics are inheritable, Darwin suggested, therein lies a mechanism for evolution. It seemed likely that some characteristics might do a better job than others in allowing the individual with them to survive and reproduce. The environment would select the better-adapted individuals, analogous to the way a breeder selects individuals with desirable traits and culls those with undesirable traits. Darwin called this process "natural selection." Continued for hundreds of generations, natural selection might produce a population so different from the original one that it could be considered a new species; and with much further evolution, it would become a new genus, then a new family, then an order, a class, and finally a phylum.

The three components of Darwin's suggestion for how evolution could occur—a finite environment, genetic variation among individuals, and natural selection—are all processes that can be studied in nature by anyone. The data that Darwin and later evolutionists obtained were all based on observations and experiments. Even before Darwin's time, geologists had observed the great thickness of the sedimentary layers of the Earth's crust and concluded that the Earth was very, very old. Today, data obtained from measurements of radioactive decay set the origin of the Earth at about 4.5 billion years ago and the origin of life about 1 billion years later. The scientific evidence does not support a young Earth, contrary to the claims of the creationists. So also the successive layers of sedimentary rocks preserve a diary of previous inhabitants in the form of the fossil record, which has been interpreted as showing the changes of species into new species. The paleontological data provide no confirmation for the fixity of the species that the creationists claim.

The usefulness of any scientific theory is its ability to explain a variety of otherwise inexplicable observations. The greater the extent to which it can do this, the more likely is it to be correct. The theory of evolution provided a naturalistic explanation for many puzzles in comparative anatomy; the classification of organisms; embryonic development; physiology; genetic makeup; the structure and physiology of cells; the biochemical reactions in microorganisms, plants, and animals; the geographic distribution of species; and observations in the paleontological record. To date, there have been no confirmed observations or experiments that falsify the theory of evolution.

Can we say, therefore, that evolution or any major concept is "true" or that it is a "fact"? Scientists prefer not to use those terms, based on their experience that all scientific "truths" and "facts" relating to major concepts have been modified and improved with time. For example, long ago it was believed that all matter was composed of four basic elements—earth, water, fire, and air. The notion that a small number of substances combine to form all other substances was a powerful idea that started in ancient Greece and has proved to be correct, but the substances first specified as elements proved to be incorrect. Later scientists hypothesized that all matter is composed of similar indivisible structures—atoms. In the nineteenth century, about 100 basic kinds of atoms had been discovered or predicted to exist, and these were thought to be the basic building blocks of all matter. Then in the twentieth century the atoms themselves were discovered to be composed of electrons, neutrons, protons, quarks, and ever smaller particles. One current theory hypothesizes that the basic building blocks of matter are superstrings—one-dimensional, vibrating massless strings only 10^{-33} centimeters long.

A useful way to look at the history of science is that the earlier statements were not wrong but merely incomplete. Each statement represented the best that could be said with the data available; when better data became available, the statements were upgraded. Nevertheless, some theories are so well established that they can be considered "true" in the common sense of that word. The theory of gravity is one of them: if a heavy object is released from the hand, it falls; we can be sure of that—as long as we remain on the Earth. In a spacecraft, heavy objects will float in air.

In the 1940s, noted English biologist Julian Huxley expressed the opinion that the theory of evolution was as well established as the theory of gravitation. No new data or experiments have changed that opinion. Nevertheless, the theory continues to be augmented with new data and interpretations, and—of the greatest importance—it suggests new lines of research and discovery. If Darwin could pay us a visit today, he would probably be delighted to find his basic ideas still useful, but he would probably be astonished at the progress made since 1859, much of it in genetics and molecular biology—fields that didn't exist during his lifetime.

PATTERNS OF THOUGHT

The differing explanations that creationists and evolutionists provide to explain life over the geological ages illustrate the two alternative ways human beings think about their world, their hopes, and their lives. One approach is rational, demanding data and logic, while the other way is more romantic, involving emotion, faith, and personal preference. The rational mode is more likely to be observed in the sciences, technology, business, law, medicine, government, and education. The romantic mode is more likely to be found in music, art, literature, religion, interpersonal relationships, and lifestyles. Most of the time both patterns of thought are involved in decision making. A desire to improve the lives of those living in poverty originates in the romantic mode, but if that desire is to bear fruit, rational thought must be applied to implementation of the goal.

The disputes between the creationists and scientists cannot be considered scientific, since only one side deals with science. Nor can they be considered religious disputes, since the theories of science have nothing to say about gods or other supernatural phenomena that cannot be studied by scientific procedures. The disputes are best thought of as political disagreements, not scientific ones. In instances where two points of view exist—each held firmly—an effort could be made by each side to maintain its own views (both will) and ignore the other.

Both sides will have to make adjustments—not in their beliefs but in their actions. The creationists can keep their faith, but they must refrain from trying to force their ideas on the science curriculum in the classroom. Professional scientists need to initiate a full discussion in our society of what science is and is not and how scientists know what they claim to know. A large percentage of Americans remains ignorant of how scientists discover new knowledge and unite that knowledge into conceptual schemes. For people not thoroughly familiar with biology and geology, the relevance of the evidence presented for evolution is not obvious without a good background in these sciences.

Little in modern science is readily understandable without considerable specialized education. Is it obvious to nonphysicists why scientists accept that atoms are made up of many different kinds of particles, how scientists know that the rotation of the Earth rather than a circling sun causes night and day, why ice floats, how the sun's energy drives the workings of our bodies, or how it was determined that DNA is the hereditary material? All of us, including scientists, are amateurs when it comes to understanding the cutting-edge developments in fields other than our own. Few biologists can comprehend discoveries in theoretical physics, and most physicists and many biologists have little understanding of the data of modern evolutionary biology. We have to accept that everyone is ignorant of almost everything.

Our acceptance of the scientists' word about so many things in nature rests on the belief that those who devote their lives to the study of a subject are more likely than others to have reliable information. And we know that when scientists reach an erroneous conclusion, the errors and omissions leading to that conclusion will eventually be detected as other scientists in the field repeat the experiments or observations and extend the analysis. In other words, the system is self-correcting, and that is its greatest strength. If equally competent scientists with equally effective equipment anywhere in the world perform the same experiments or make the same kinds of observations, the results are likely to be similar. If the outcomes differ, repetitions of the work will reveal the sources of the differences. Science advances by the retention of the provable and the elimination of the falsifiable. Though politically and culturally we are far from seeing ourselves as one world, scientifically we come very close.

Professional scientists need to become informed enough to discuss, as citizens repeatedly ask them to, the difference between creationism based on faith and evolution based on confirmable evidence. And religious leaders should support the scientists in this effort. Many theologians deplore the erroneous statements and the political activities of creationists. They could make clear to their congregations that acceptance of evolution does not mean denial of religion, but it does mean regarding the P version of creation as metaphor, not science. And they should inform their parishioners that many of the major religions in the Western world and their leaders accept evolution as the best explanation for the organic diversity of the present and the past.

TEACHING SCIENCE

Concerned parents and other citizens should do their part by working with school boards and teachers to make sure that science is taught as it has been developed by scientists and not according to the views of those who wish to discount it. The educational system, which involves parents, schools, and society, does not give adequate consideration to developing both the romantic and rational aspects of the mind. The world of young children has a rich romantic and supernatural component, with influences from children's books, motion pictures, churches, and parents. Little consideration is given to the processes that require evidence and critical thought for reaching conclusions. Much anguish and conflict might be avoided later if children were taught in their formative years that both romantic and rational thought patterns are valuable for effective interrelations of the individuals of any society.

Children in kindergarten and elementary school are highly receptive to topics in science, especially if those topics pertain to the world they know. Through simple observations and experiments, children learn that that they can answer some questions about nature themselves. Unfortunately, teachers in these beginning grades rarely are knowledgeable enough about the natural sciences and critical thinking to lead their students in explorations and analyses that will help develop their rational patterns of thought. Correcting this deficiency will require a major overhaul of the existing educational system—both in the ways teachers are educated and in the opportunities for children to learn.

Few colleges and universities offer science courses specifically designed for prospective schoolteachers. Science courses in higher education are designed primarily for students who plan to be professional scientists, engineers, or physicians. The curriculum for these students tends to be rigorous, selective, and restrictive. But such courses alone do not provide an adequate preparation for teachers. Prospective teachers should have a solid background in the subjects they will teach, but they also need courses to help them teach science in ways appropriate for elementary schools, middle schools, or high schools. They must be familiar with all fields of science and how they relate to the lives of human beings. Students should learn not only the basic principles of all sciences but also topics such as the differences between science and other disciplines. Critical thinking and what counts as evidence should be understood. There should always be a deep concern for the relationship between the sciences and public policy and the conditions of the environment that determine human welfare. And for all levels of education, and appropriate for the ages of the students, science should consider the concerns and interests of people.

Many high school science teachers actually teach creationism, which can only be regarded as a lack of professionalism and a violation of law. But this may not always be the teacher's fault. Few college and university science courses deal with the creation-versus-evolution controversy explicitly, such that students, including those who will be teachers, can understand the issues and the data and be able to make sound judgments. Again, scientists must accept much of the responsibility for this situation.

MOVING FORWARD

The evolution-versus-creationism debate can be viewed as a conflict between two incompatible paradigms, the romantic and the rational. For centuries humankind has spoken of the affairs of the head and of the heart as different realms. Romantically they are; rationally they are not. Both scientists and humanists require both a thinking head and a pumping heart. The prognosis for those who choose one or the other is poor. The challenge is to employ the metaphorical heart and the metaphorical head in ways that emphasize the strengths and limitations of each.

The feeling of awe that comes from understanding the natural world and the unifying principles that control the interactions of all matter can be a deep religious experience. One need only think about the long history of the great mountain ranges to be awed by the mighty forces of the crustal plates crashing against one another, pushing the surface upward to form mountains, and the power of the wind and rain that sculpted them. And every living creature reminds one not only of its long history going back to the origin of life itself but of the extraordinary structures and functions that permit its survival in what can be an unkind world. One can experience a profound sense of wonder in being a part of a knowable world.

Whether or not God is the driving force behind that world is not for any scientist, speaking as a scientist, to say. Scientists as scientists can deal only with phenomena that can be studied, and this does not include God. As individuals, however, they can accept that there is a God—a position that is based on belief, not scientific evidence. According to author and editor Gregg Easterbrook (1997), there is a growing feeling among many scientists and religious leaders that accommodation and even mutual support of their positions may be the mode of the future: "Perhaps the fact that the two schools of thought have so often been at each other's throats stems from mutual recognition of their linked destinies, and their joint commitment to the idea that the truth is out there. Rather than being driven ever farther apart, tomorrow's scientist and theologian may seek each other's solace" (693). Harvard biologist E. O. Wilson (1998) also predicts that science and religion are moving toward synthesis.

Physicist, cosmologist, and Nobel laureate Charles H. Townes expands upon this point of view (1995):

> The ever-increasing success of science has posed many challenges and conflicts for religion—conflicts which are resolved in individual lives in a variety of ways. Some accept both religion and science as dealing with quite different matters by different methods, and thus separate them so widely in their thinking that no direct confrontation is possible. Some repair rather completely to the camp of science or of religion and regard the other as ultimately of little importance, if not downright harmful. To me science and religion are both universal, and basically very similar. In fact, to make the argument clear, I should like to adopt the rather extreme point of view that their differences are largely superficial, and that the two become almost indistinguishable if we look at the real nature of each. . . . (157)
>
> The goal of science is to discover the order in the universe, and to understand through it the things we sense around us, and even man himself. This order we express as scientific principles or laws, striving to state them in the simplest and yet most inclusive ways. The goal of religion may be stated, I believe, as an understanding (and hence acceptance) of the purpose and meaning of our universe and how we fit into it. Most religions see a unifying and inclusive origin of meaning, and this purposeful force we call God. . . .
>
> The essential role of faith in religion is so well known that it is usually taken as characteristic of religion, and as distinguishing religion from science. But faith is essential to science too, although we do not so generally recognize the basic need and nature of faith in science.
>
> Faith is necessary for the scientist to even get started, and deep

faith necessary for him to carry out his tougher tasks. Why? Because he must be personally committed to the belief that there is order in the universe and that the human mind—in fact his own mind—has a good chance of understanding this order. Without this belief, there would be little point in intense effort to try to understand a presumably disorderly or incomprehensible world. Such a world would take us back to the days of superstition, when man thought capricious forces manipulated his universe. In fact, it is just this faith in an orderly universe, understandable to man, which allowed the basic change from an age of superstition to an age of science, and has made possible our scientific progress. . . . (161–62)

Finally, if science and religion are so broadly similar, and not arbitrarily limited in their domains, they should at some time clearly converge. I believe this confluence is inevitable. For they both represent man's efforts to understand his universe and must ultimately be dealing with the same substance. . . . But converge they must, and through this should come new strength for both. (166)

It is hard to imagine a synthesis occurring anytime in the near future because science and religion still use very different thought patterns—one based on evidence and the other on belief. Rather than predicting a synthesis of science and religion, it may be more useful to regard them as coequals with different domains. Science attempts to understand how the natural world works, and this knowledge not only feeds our natural curiosity but also provides us with astonishing power to live and to change our lives. But for most people this knowledge is not enough; the question "What does it all mean?" is equally important. In a general way we can say that religion deals with questions of meaning as well as with standards of conduct. But these meanings and standards are assigned by human beings, not derived from what science tells us about the mechanisms of life. There is one science for the entire world, but innumerable and often conflicting answers to questions about the purpose and meaning of things. "What is the purpose of human life?" has answers that vary from one human being to another and even vary at different times for the same individual. Herein lies a gap that is so far unbridged. Whereas science is discovered, meaning is assigned.

From this standpoint, science and religion need not be seen in conflict because their domains are distinct. We should recognize that the first is preeminent in providing knowledge of the natural world that can be exciting and awe-inspiring as well as provide for modern medicine and the comforts of modern life. Those comforts are not to be enjoyed unless each society can agree on a moral code of behavior so necessary for civilized life.

However, science and creationism come into serious conflict when the politically active antievolution creationists, or creation scientists as they sought to be called until recently, campaign to have their brand of creationism taught in the schools alongside evolution or to be given "equal time." This subset of creationists are not scientists; they do not see new information in laboratory studies or field explorations to test their hypothesis. They ignore the conclusions of scientists who over the past two centuries have provided a body of confirmable evidence that leads to the conclusion that, beyond all reasonable doubt, evolution has been the dominant phenomenon in life over the ages. Creationists have every right to ignore all this and believe whatever suits them, but as the highest courts of the nation and some states have ruled, creationism is not science and cannot be taught as science in the nation's public schools.

Religion has had a checkered influence during human history; it has been responsible for much bloodshed and misery in the past as well as today in many parts of the world. This is less a function of religion itself and more the interpretation of religion as sanctioning what individuals, cultures, and nations wish to do. Battling nations each proclaim that God is on their side, which suggests a robust polytheism. "Go ye into all the world and preach the gospel to every nation" has been read as a command to destroy the cultures of many native societies in North and South America, Australia, sub-Saharan Africa, and islands throughout the world. And much of this strife among different peoples has been supported by discoveries in science and the conversion of that knowledge into instruments of personal and mass destruction. Science as a way of knowing nature is neutral, but its power can be used for infinite good or infinite evil. The choice is ours.

Throughout the ages there have always been a few who, without abandoning their religion, did forgo the blind acceptance of dogma and superstition. They undertook the extraordinarily difficult, lonely, and frequently dangerous path of using their unfettered minds for rational inquiry. It is these individuals who have given us the modern world and the possibility of truly great improvement of the human condition. They have replaced the primitive view of nature as chaotic, mysterious, and often threatening with a view of the universe and life as responding in patterns that are precise, beautiful, and awe-inspiring. Beyond giving pleasure to the inquisitive, analytical mind, this progress in understanding provides previously unimagined ways to feed the hungry, heal the sick, and lessen toil. Lives are poorly lived when they look out upon a cold, hostile, inscrutable world; lives are enhanced when they look out upon a world with an appreciation of its beauty and order and its suitability as a warm and friendly home. It matters little for the great moral and ethical questions facing humanity whether or not the human brain and mind are consequences of random events in evolution, though scientists are convinced they are. However, it matters a great deal that we use our brains and minds honestly, humanely, intensively, and effectively to preserve and improve the world for ourselves and for the generations that follow.

SUGGESTED READING

SCIENCE AND RELIGION

Draper (1874) and White (1898) are classic nineteenth-century accounts. More recent, and more optimistic, treatments are Ahlstrom (1972), Barbour (1966), Birch (1995), Davies (1992), Dawkins (1986, 1998), Dillenberger (1960), Gilbert (1997), Gilkey (1959), Gillespie (1979), Godfrey (1983), Goldberg (1999), Goodenough (1998), Gould (1999), Greene (1959a), Hooykaas (1972), W. T. Jones (1965), Lindberg and Numbers (1986), Livingston (1987), Miller (1999), J. R. Moore (1979), Pennock (1999), Polkinghorne (1998), Raymo (1987, 1998), Roberts (1988), Russell (1961), Russett (1976), J. A. Thomson (1925), Tourney (1994), Townes (1995), and Ward (1996).

HISTORY OF THE ANCIENT NEAR EAST

Buttrick (1952), Eliade (1978), C. M. Jones (1971), and Thompson (1986).

THE JUDEO-CHRISTIAN BIBLE

Alter (1996), Asimov (1968), Davidson (1973), Friedman (1987, 1998), Fretheim (1994), Harris (1997), Kugel (1997), Lace (1972), Mellor (1972), Mitchell (1996), Pritchard (1969), Richardson (1953), and Thompson (1986). In Buttrick

(1952) are sections on the background of the Old Testament as well as a scholarly discussion of Genesis. A new edition, *The New Interpreter's Bible,* is now under way. The first volume, containing the background chapters and Genesis, was published in 1994.

RELIGION AND MYTHOLOGY

Campbell (1962, 1964), Doria and Lenowitz (1976), Eliade (1959), Graves (1955), Heidel (1951), James (1978), Kirk (1973), Lemming and Lemming (1994), Long (1963), Sandars (1964), and Torrance (1994).

DARWINISM AND EVOLUTION

Barlow (1958), Browne (1996), Darwin (1859), Dawkins (1986, 1996), Desmond (1997), Desmond and Moore (1991), Eldridge (1999), Flew (1997), Fortey (1998), Futuyma (1983, 1998), Greene (1959b), Irvine (1955), Lewin (1996), Mayr (1991), J. A. Moore (1993), National Academy of Sciences (1998, 1999), Numbers (1998), Ridley (1996), Rose (1998), Schopf (1999), Strickberger (1996), Weiner (1994), Whitfield (1993), and R. J. Wilson (1967). The books by Dawkins, Greene, National Academy of Sciences, Weiner, and Whitfield are good introductions. Eldridge, Futuyma (1998), Ridley, and Strickberger are comprehensive college-level textbooks. Darwin (1859) is a facsimile of the first edition of the *Origin* and is especially valuable for its introduction by Ernst Mayr.

HUMAN EVOLUTION

Johanson and Edgar (1996), Jones, Martin, and Pilbeam (1992), Lewin (1984), and Tattersall (1993, 1995).

FUNDAMENTALISM AND CREATIONISM

Two fine introductions to the creationists and their programs are Numbers (1992) and, for more recent developments, Pennock (1999). Other good sources are Berra (1990), Eve and Harrold (1991), Futuyma (1983), Godfrey

(1983), Greene (1959a), Holton (1993), Kaiser (1997), Kitcher (1982), Larson (1989), Marsden (1980), Marty and Applebey (1993), J. A. Moore (1982, 1983), Nelkin (1977), Newell (1982), Pennock (1999), Shermer (1997, 2000), and Strahler (1987). Two basic books by creationists that describe their beliefs are Morris (1974) and Whitcomb and Morris (1961). In addition, the Creation Research Society has issued a biology textbook that promotes creationism and argues against evolution: J. N. Moore and H. S. Slusher (1970). Additional information can be sought from the Institute for Creation Research, P.O. Box 2666, El Cajon, CA 92021-9982. Recent works by the new creationists of the intelligent design persuasion are Austin (1994), Behe (1996), Johnson (1991, 1997), and Dembski (1998). An organization that has been responding to the creationists' attacks on evolution and their attempts to have creationism taught in the schools and that can supply information is the National Center for Science Education, P.O. Box 9477, Berkeley, CA 94709-0477. Its *Reports of the National Center for Science Education* is published six times a year and contains articles relating to creationism and evolution.

THE SCOPES TRIAL

Allen (1967), de Camp (1968), Ginger (1958), Larson (1997), Lawrence and Lee (1955, a play based on the trial), Scopes and Presley (1967), Scopes Trial (1925, court transcript of the trial), Shipley (1927), Tierney (1979), Tompkins (1965), and Weinberg and Weinberg (1980).

THE FLIGHT FROM SCIENCE AND REASON

Recent titles on this very important and divisive subject are Dawkins (1998), Gross and Levitt (1994), Gross, Levitt, and Lewis (1997), Harrold and Eve (1995), Holton (1993), Koertge (1998), Sagan (1996), Shermer (1997), and Sokal and Bricmont (1998).

REFERENCES

Agassiz, Louis. 1860. Professor Agassiz on the Origin of Species. *American Journal of Science* (2nd ser.) 30: 142–54.

Ahlstrom, Sidney E. 1972. *A Religious History of the American People.* New Haven: Yale University Press.

Allen, Leslie H. 1967. *Bryan and Darrow at Dayton: The Record and Documents of the Bible-Evolution Trial.* New York: Russell and Russell.

Alter, Robert. 1996. *Genesis: Translation and Commentary.* New York: Norton.

Asimov, Isaac. 1968. *Asimov's Guide to the Bible: The Old Testament.* New York: Avon.

Austin, S. A., ed. 1994. *Grand Canyon: Monument to Catastrophe.* Santee, Calif.: Institute for Creation Research.

Barbour, Ian G. 1966. *Issues in Science and Religion.* New York: Prentice-Hall.

Barlow, Nora. 1958. *The Autobiography of Charles Darwin, 1809–1882.* London: Collins.

Behe, Michael J. 1996. *Darwin's Black Box: The Biochemical Challenge to Evolution.* New York: Free Press.

Berra, Tim M. 1990. *Evolution and the Myth of Creationism: A Basic Guide to the Facts in the Evolution Debate.* Stanford: Stanford University Press.

Birch, Charles. 1995. *Feelings.* Sydney: New South Wales Press.

Bowler, Peter. 1989. *Evolution: The History of an Idea.* Berkeley: University of California Press.

Browne, Janet. 1996. *Charles Darwin: Voyaging.* Princeton: Princeton University Press.

Bryan, W. J. 1923. The Fundamentals. *The Forum* 70: 1665–80.

Buttrick, G. A., ed. 1952. *The Interpreter's Bible.* Vol. 1. New York: Abingdon Press.

Campbell, Joseph. 1962. *The Masks of God: Oriental Mythology.* New York: Viking.

———. 1964 *The Masks of God: Occidental Mythology.* New York: Viking.

Crim, Keith R. 1994. Modern English Versions of the Bible. In *The New Interpreter's Bible,* vol. 1. Pp. 22–32. New York: Abingdon Press.

Darwin, Charles. 1859. *On the Origin of Species by Natural Selection, or the Preservation of Favoured Races in the Struggle for Life.* Rpt., Cambridge: Harvard University Press, 1964.

Darwin, Charles, and Alfred Russel Wallace. 1958. *Evolution by Natural Selection.* Cambridge: Cambridge University Press.

Darwin, Francis. 1888. *Life and Letters of Charles Darwin.* 3 vols. London: John Murray.

Davidson, Robert. 1973. *Genesis 1–11. The Cambridge Bible Commentary on the New English Bible.* Cambridge: Cambridge University Press.

Davies, Paul. 1992. *The Mind of God.* New York: Simon and Schuster.

Dawkins, Richard. 1986. *The Blind Watchmaker.* New York: Norton.

———. 1996. *Climbing Mount Improbable.* New York: Norton.

———. 1998. *Unweaving the Rainbow: Science, Delusion, and the Appetite for Wonder.* Boston: Houghton Mifflin.

de Camp, L. Sprague. 1968. *The Great Monkey Trial.* Garden City, N.Y.: Doubleday.

Dembski, William A. 1998. *The Design Inference.* New York: Cambridge University Press.

Desmond, Adrian. 1997. *Huxley: From Devil's Disciple to Evolution's High Priest.* Reading, Mass.: Addison-Wesley.

Desmond, Adrian, and James Moore. 1991. *Darwin.* New York: Warner.

Diamond, Jared. 1992. *The Third Chimpanzee: The Evolution and Future of the Human Animal.* New York: Harper-Collins.

di Gregorio, Mario A. 1984. *T. H. Huxley's Place in Natural Science.* New Haven: Yale University Press.

Dillenberger, John. 1960. *Protestant Thought and Natural Science.* Nashville: Abingdon Press.

Doria, Charles, and Harris Lenowitz. 1976. *Origins: Creation Texts from the Ancient Mediterranean.* Garden City, N.Y.: Anchor Books.

Draper, John William. 1874. *History of the Conflict between Religion and Science.* New York: Appleton.

Easterbrook, Gregg. 1997. Science and God: A Warming Trend? *Science* 277: 890–93.

Elders, Wilfred A. 1998. Bibliolatry in the Grand Canyon. *Reports of the National Center for Science Education* 18: 8–15.

Eldridge, Niles. 1999. *The Pattern of Evolution.* New York: W. H. Freeman.

Eliade, Mircea. 1959. *The Sacred and the Profane: The Nature of Religion.* New York: Harcourt Brace Jovanovich.

———. 1978. *A History of Religious Ideas.* Vol. 1, *From the Stone Age to the Eleusinian Mysteries.* Chicago: University of Chicago Press.

Eve, Raymond, and Francis B. Harrold. 1991. *The Creationist Movement in Modern America.* Boston: Twayne.

Flew, Antony. 1997. *Darwin Evolution.* 2nd ed. London: Transaction.

Fortey, Richard A. 1998. *A Natural History of the First Four Billion Years.* New York: Knopf.

Fretheim, Terence E. 1994. Genesis. In *The New Interpreter's Bible,* vol. 1. Nashville: Abingdon Press.

Friedman, Richard E. 1987. *Who Wrote the Bible?* New York: Summit Books.

———. 1998. *The Hidden Book in the Bible: Restored, Translated, and Introduced.* San Francisco: Harper.

Futuyma, Douglas J. 1983. *Science on Trial: The Case for Evolution.* New York: Pantheon.

———. 1998. *Evolutionary Biology.* 3rd ed. Sunderland, Mass.: Sinauer.

Gilbert, James. 1997. *Redeeming Culture: American Religion in an Age of Science.* Chicago: University of Chicago Press.

Gilkey, Langdon. 1959. *Maker of Heaven and Earth.* Garden City, N.Y.: Doubleday.

Gillespie, Neal C. 1979. *Charles Darwin and the Problem of Creation.* Chicago: University of Chicago Press.

Ginger, Ray. 1958. *Six Days or Forever: Tennessee v. John Thomas Scopes.* Boston: Beacon Press.

Godfrey, Laurie R., ed. 1983. *Scientists Confront Creationism.* New York: Norton.

Goldberg, Steven. 1999. *Seduced by Science: How American Religion Has Lost Its Way.* New York: New York University Press.

Goodenough, Ursula. 1998. *The Sacred Depths of Nature.* New York: Oxford University Press.

Gould, Stephen J. 1999. *Rocks of Ages: Science and Religion in the Fullness of Life.* New York: Ballantine.

Graves, Robert. 1955. *The Greek Myths.* 2 vols. Baltimore: Penguin Books.

Gray, Asa. 1860. Review of Darwin's Theory on the Origin of Species by Means of Natural Selection. *American Journal of Science* (2nd ser.) 29: 153–84.

Gray, Thomas. 1984. University Course Reduces Belief in Paranormal. *Skeptical Inquirer* 8: 247–51.

Greene, John C. 1959a. *The Death of Adam: Evolution and Its Impact on Western Thought.* Ames: Iowa State University Press.

———. 1959b. *Darwin and the Modern World View.* Baton Rouge: Louisiana State University Press.

Gross, Paul R., and Norman Levitt. 1994. *Higher Superstition: The Academic Left and Its Quarrels with Science.* Baltimore: Johns Hopkins University Press.

Gross, Paul R., Norman Levitt, and Martin W. Lewis, eds. 1997. *The Flight from Science and Reason.* Baltimore: Johns Hopkins University Press.

Harris, Stephen L. 1997. *Understanding the Bible.* 4th ed. London: Mayfield.

Harrold, Francis B., and Raymond A. Eve, eds. 1995. *Cult Archaeology and Creationism.* Iowa City: University of Iowa Press.

Heidel, Alexander. 1951. *The Babylonian Genesis.* Chicago: University of Chicago Press.

Herodotus. 1987. *The History.* Chicago: University of Chicago Press.

Holton, Gerald. 1993. *Science and Anti-Science.* Cambridge: Harvard University Press.

Hooykaas, R. 1972. *Religion and the Rise of Modern Science.* Edinburgh: Scottish Academic Press.

Huxley, Thomas H. 1863a. *Evidence as to Man's Place in Nature.* New York: Appleton.

———. 1863b. *On the Origin of Species: or, The Causes of the Phenomena of Organic Nature: A Course of Six Lectures to Working Men.* New York: Appleton.

———. 1877. *American Addresses, with a Lecture on the Study of Biology.* London: Macmillan.

———. 1878. *A Manual of the Invertebrated Animals.* New York: Appleton.

Irvine, William. 1955. *Apes, Angels, and Victorians: The Story of Darwin, Huxley, and Evolution.* New York: McGraw-Hill.

James, William. 1978. *The Varieties of Religious Experience: Being the Gifford Lectures on Natural Religion Delivered at Edinburgh in 1901–1902.* Garden City, N.Y.: Doubleday.

Johanson, Donald, and Blake Edgar. 1996. *From Lucy to Language.* New York: Simon and Schuster.

John Paul II. 1997. The Pope's Message on Evolution and Four Commentaries. *Quarterly Review of Biology* 72: 381–406.

Johnson, Phillip. 1993. *Darwin on Trial.* 2d ed. Lanham, Md.: Regnery Gateway.

———. 1997. *Defeating Darwinism by Opening Minds.* Downers Grove, Ill.: Inter-Varsity Press.

Jones, Clifford M. 1971. *Old Testament Illustrations: The Cambridge Bible Commentary.* Cambridge: Cambridge University Press.

Jones, Steve, Robert Martin, and David Pilbeam, eds. 1992. *Cambridge Encyclopedia of Human Evolution.* New York: Cambridge University Press.

Jones, W. T. 1965. *The Sciences and the Humanities.* Berkeley: University of California Press.

Kaiser, Christopher B. 1997. *Creational Theology and the History of Physical Science: The Creationist Tradition from Basil to Bohr.* New York: Brill.

Kirk, G. S. 1973. *Myth: Its Meanings and Functions in Ancient and Other Cultures.* Berkeley: University of California Press.

Kitcher, Philip. 1982. *Abusing Science: The Case against Creationism.* Cambridge: MIT Press.

Koertge, Noretta. 1998. *A House Built on Sand: Exposing Postmodernist Myths about Science.* New York: Oxford University Press.

Kugel, James L. 1997. *The Bible as It Was.* Cambridge: Harvard University Press.

Lace, O. Jessie, ed. 1972. *Understanding the Old Testament: The Cambridge Bible Commentary.* Cambridge: Cambridge University Press.

Larson, Edward. J. 1989. *Trial and Error: The American Controversy over Creation and Evolution.* New York: Oxford University Press.

————. 1997. *Summer for the Gods: The Scopes Trial and America's Continuing Debate over Science and Religion.* New York: Basic Books.

Lawrence, Jerome, and Robert E. Lee. 1955. *Inherit the Wind.* New York: Random House.

Lawson, Anton E., and John Weser. 1990. The Rejection of Nonscientific Beliefs about Life: Effects of Instruction and Reasoning Skills. *Journal of Research in Science Teaching* 27: 589–606.

Lemming, David, and Margaret Lemming. 1994. *Creation Myths.* New York: Oxford University Press.

Lewin, Roger. 1984. *Human Evolution: An Illustrated Introduction.* New York: W. H. Freeman

————. 1996. *Patterns of Evolution.* New York: Scientific American Library.

Lindberg, David C., and Ronald L. Numbers, eds. 1986. *God and Nature: Historical Essays on the Encounter between Christianity and Science.* Berkeley: University of California Press.

Livingston, David N. 1987. *Darwin's Forgotten Defenders: The Encounter between Evangelical Theology and Evolutionary Thought.* Grand Rapids, Mich.: William B. Eerdmans.

Long, Charles H. 1963. *Alpha: The Myths of Creation.* New York: Collier Books.

Macintosh, A. A. 1972. From the Ancient Languages to the New English Bible. In *The Cambridge Bible Commentary on the New English Bible: The Making of the Old Testament,* edited by Enid B. Mellor. Pp. 133–66. Cambridge: Cambridge University Press.

Marsden, George M. 1980. *Fundamentalism and American Culture: The Shaping of Twentieth-Century Evangelicalism, 1870–1925.* New York: Oxford University Press.

Marsh, O. C. 1874. Fossil Horses in America. *American Naturalist* 8: 288–301.

————. 1880. *Odontornithes: Monograph of the Extinct Toothed Birds of North America.* United States Geological Exploration of the Fortieth Parallel. Vol. 7. Washington, D.C.: U.S. Government Printing Office.

Marty, M. E., and R. S. Applebey, eds. 1993. *Fundamentalism and Society: Reclaiming the Sciences, the Family, and Education.* Chicago: University of Chicago Press.

Matsumura, Molleen, ed. 1995. *Voices for Evolution.* Rev. ed. Berkeley: National Center for Science Education.

Mayr, Ernst. 1991. *One Long Argument: Charles Darwin and the Genesis of Modern Evolutionary Thought.* Cambridge: Harvard University Press.

Mellor, Enid B., ed. 1972. *The Making of the Old Testament.* Cambridge: Cambridge University Press.

Miller, Kenneth. 1999. *Finding Darwin's God: A Scientist's Search for Common Ground between God and Evolution.* New York: Cliff Street Books.

Mitchell, Stephen. 1996. *Genesis: A New Translation of the Classical Biblical Stories.* New York: Harper-Collins.

Montague, Ashley, ed. 1984. *Science and Creationism.* New York: Oxford University Press.

Moore, James R. 1979. *The Post-Darwinian Controversies: A Study of the Protestant Struggle to Come to Terms with Darwin in Great Britain and America, 1870–1900.* Cambridge: Cambridge University Press.

Moore, John A. 1982. Evolution and Public Education. *Bioscience* 32: 606–12.

———. 1983. Why Are There Creationists? *Journal of Geological Education* 31: 95–104.

———. 1993. *Science as a Way of Knowing.* Cambridge: Harvard University Press.

Moore, John N., and H. S. Slusher. 1970. *Biology: A Search for Order in Complexity.* Grand Rapids, Mich.: Zondervan Publishing House.

Morris, Henry M. 1974. *Scientific Creationism.* San Diego: Creation-Life Publishers.

National Academy of Sciences. 1998. *Teaching about Evolution and the Nature of Science.* Washington, D.C.: National Academy Press.

———. 1999. *Science and Creationism: A View from the National Academy of Sciences.* 2nd ed. Washington, D.C.: National Academy Press.

Nelkin, Dorothy. 1977. *Science Textbook Controversies and the Politics of Equal Time.* Cambridge: MIT Press.

Newell, Norman D. 1982. *Creation and Evolution: Myth or Reality?* New York: Columbia University Press.

Numbers, Ronald L. 1992. *The Creationist: The Evolution of Scientific Creationism.* New York: Knopf.

———. 1998. *Darwinism Comes to America.* Cambridge: Harvard University Press.

Orlinsky, Harry M., ed. 1969. *Notes on the New Translation of the Torah.* Philadelphia: Jewish Publication Society.

Overton, William R. 1982. Creationism in Schools: The Decision in *McLean versus the Arkansas Board of Education*. *Science* 215: 934–43.

Paley, William. 1811. *Natural Theology; or, Evidences of the Existence and Attributes of the Deity.* 13th edition. London: J. Faulder et al. 1st ed. 1802.

Pennock, Robert T. 1999. *Tower of Babel: The Evidence against the New Creationism.* Cambridge: MIT Press.

Polkinghorne, John. 1998. *Belief in God in an Age of Science.* New Haven: Yale University Press.

Power, Henry. 1664. *Experimental Philosophy.* Rpt., New York: Johnson, 1966.

Preston, Douglas. 1997. The Lost Man. *New Yorker.* (June 16) 70–81.

Pritchard, James B., ed. 1969. *Ancient Near East Texts Relating to the Old Testament.* 3rd ed. Princeton: Princeton University Press.

Radcliffe-Brown, A. R. 1933. *The Andaman Islanders.* Rpt., Glencoe, Ill.: Free Press, 1948.

Raymo, Chet. 1987. *Honey from Stone: A Naturalist's Search for God.* New York: Dodd Mead.

———. 1998. *Skeptics and True Believers: The Exhilarating Connection between Science and Religion.* New York: Walker.

Richardson, Alan. 1953. *Genesis 1–11.* London: SCM Press.

Ridley, Mark. 1996. *Evolution.* Boston: Blackwell Scientific.

Roberts, J. H. 1988. *Darwinism and the Divine in America: Protestant Intellectuals and Organic Evolution, 1859–1900.* Madison: University of Wisconsin Press.

Rose, Michael R. 1998. *Darwin's Spectre.* Princeton: Princeton University Press.

Russell, Bertrand. 1961. *Religion and Science.* New York: Oxford University Press.

Russett, Cynthia Eagle. 1976. *Darwin in America: The Intellectual Response, 1865–1912.* San Francisco: W. H. Freeman.

Sagan, Carl. 1996. *The Demon-Haunted World: Science as a Candle in the Dark.* New York: Ballantine Books.

Sandars, N. K. 1964. *The Epic of Gilgamesh.* Baltimore: Penguin Books.

Schopf, J. William. 1999. *Cradle of Life: The Discovery of the Earth's Earliest Fossils.* Princeton: Princeton University Press.

Scopes, John T., and James Presley. 1967. *Center of the Storm: Memoirs of John T. Scopes.* New York: Holt, Rinehart, and Winston.

[Scopes Trial.] 1925. *The World's Most Famous Court Trial: Tennessee Evolution Case.* Cincinnati: National Books Company.

Shapin, Steven. 1998. *The Scientific Revolution.* Chicago: University of Chicago Press.

Shermer, Michael. 1997. *Why People Believe in Weird Things: Pseudoscience, Superstition, and Other Confusions of Our Time.* New York: W. H. Freeman.

———. 2000. *How We Believe: The Search for God in an Age of Science.* New York: W. H. Freeman.

Shipley, Maynard. 1927. *The War on Modern Science: A Short History of the Fundamentalist Attacks on Evolution and Modernism.* New York: Knopf.

Sibley, C. G., and J. E. Ahlquist. 1987. DNA Hybridization Evidence of Hominoid Phylogeny: Results from an Expanded Data Set. *Journal of Molecular Evolution* 26: 99–121.

Singer, Barry, and Victor A. Benassi. 1981. Occult Beliefs. *American Scientist* 69: 49–55.

Sokal, Alan, and Jean Bricmont. 1998. *Fashionable Nonsense: Postmodern Intellectuals' Abuse of Science.* New York: Picador.

Springer, Mark S., et al. 1997. Endemic African Mammals Shake the Phylogenetic Tree. *Nature* 388: 61–64.

Strahler, Arthur N. 1987. *Science and Earth History: The Evolution/Creation Controversy.* Buffalo: Prometheus Books.

Strickberger, Monroe. 1996. *Evolution.* 2nd ed. Boston: Jones and Barlett.

Tattersall, Ian. 1993. *The Human Odyssey: Four Million Years of Human Evolution.* New York: Prentice Hall.

———. 1995. *The Fossil Trail: How We Know What We Think We Know about Human Evolution.* New York: Oxford University Press.

Thompson, J. A. 1986. *Handbook of Life in Bible Times.* Downers Grove, Ill.: Inter-Varsity Press.

Thomson, J. Arthur. 1925. *Science and Religion: Being the Morse Lectures for 1924.* New York: Charles Scribner's.

Thomson, Keith Stewart. 1997. Natural Theology. *American Scientist* (May–June) 219–21.

Tierney, Kevin. 1979. *Darrow: A Biography.* New York: Thomas Y. Crowell.

Tompkins, Jerry R., ed. 1965. *D-day at Dayton: Reflections on the Scopes Trial.* Baton Rouge: Louisiana State University Press.

Torrance, Robert M. 1994. *The Spiritual Quest: Transcendence in Myth, Religion, and Science.* Berkeley: University of California Press.

Tourney, Christopher P. 1994. *God's Own Scientists: Creationists in a Secular World.* New Brunswick: Rutgers University Press.

Townes, Charles H. 1995. *Making Waves.* Woodbury, N.Y.: American Institute of Physics.

United States Reports. 1990. Volume 482. *Cases Adjudged in the Supreme Court at the October Term, 1986.* Washington, D.C.: U.S. Government Printing Office.

Ward, Keith. 1996. *Religion and Creation.* New York: Oxford University Press.

Weinberg, Arthur, and Lila Weinberg. 1980. *Clarence Darrow: A Sentimental Rebel.* New York: Putnam.

Weiner, Jonathan. 1994. *The Beak of the Finch: A Story of Evolution in Our Time.* New York: Knopf.

Whitcomb, John C., Jr., and Henry M. Morris. 1961. *The Genesis Flood: The Biblical Record and Its Scientific Implications.* Philadelphia: The Presbyterian and Reformed Publishing Co.

White, Andrew Dickson. 1898. *A History of the Warfare of Science with Theology in Christendom.* 2 vols. New York: Appleton.

Whitfield, Philip. 1993. *From So Simple a Beginning.* New York: Macmillan.

Wilson, E. O. 1998. *Consilience: The Unity of Knowledge.* New York: Knopf.

Wilson, R. J., ed. 1967. *Darwinism and the American Intellectual: A Book of Readings.* Homewood, Ill.: Dorsey Press.

Wise, D. U. 1998. Creationist's Geological Time Scale. *American Scientist* 86: 160–63.

INDEX

Text:	11/15 Granjon
Display:	Granjon
Compositor:	Binghamton Valley Composition, LLC
Printer/binder:	Sheridan Books, Inc.